T0149412

While living on Creek road I began searching for 'me'. I began to question what I wanted from life and just where life is taking me. I wrote down three questions I had to find answers to: What did you love as a child, what would you do if you couldn't fail at it and who do you think you are? I found a photo of myself when I was eighteen months old, very cute if I do say so dressed up in pretty pink dress and smiling for the studio photographer. To me this was a moment caught in time reflecting my authentic 'self', unshaped, free willed and without fear. From that space I concluded I was influenced for better or worse by my surroundings and people such as teachers, parents and later the media and television. I conformed to how society wanted me to be and being a female my expected role was to be a wife and mother and wherever my husband went it was my duty to follow. I had spent many years when my children and husband came first but now it was my turn. I began to read a lot of books and I was 'drawn' to self-healing and personal development. I was looking for a direction in books like Weekend Confidence Coach, How to Lose Friends and Infuriate People, Feelings Buried Alive Never Die and You Can Heal Your Life. I had an insatiable thirst for books that in later years introduced me to the unexplained and the spirits of our natural world. I discovered many inspirational affirmations and the one that made an immediate impression was 'do what you do best but do it better'. I suddenly realized there was no reason for me to continue searching for a new 'me' or a new direction, farming and the land was my life and it was what I knew best. I loved my job and couldn't see myself doing anything else, but I would need to do it better, a whole lot better!

In the back of my mind I had the idea of farming organically. Fonterra was offering a premium for the first three years while converting to organics and I liked the idea of doing my bit for the environment. In our home we were already changing our toiletries and grocery items for certified organic brands. My daughter and I were reading the ingredient lists and looking them up in our little red 'Chemical Maze' book. There were a lot we couldn't even read and the little red book had them listed as possibly dangerous to our health. I was told if you can't pronounce their names they're probably not good for you.

On the farm we were searching for something to improve the cow's health and to increase their chances of getting in calf. I stumbled upon AgriSea, an animal health tonic made from seaweed containing all the minerals the cows needed. As I read the list of vitamins, minerals and amino acids in the mixture I noticed it had everything that we had been giving them, copper, selenium, calcium, magnesium plus more. We were amazed by how fast the herd's health turned around. Their coats started to shine and they were a lot happier. I no longer had to give them a copper injection or the selenium drench, it was all in the animal tonic in a naturally balanced form. The girls loved it, even the young stock who would fight to get their share from the drench gun as we stood amongst them in the paddock!

We were beginning to change our ways and the more we looked the more flaws we found in the industrial farming methods we had been implementing for many years.

You Have Been Given a Gift

JANETTE PERRETT

BALBOA
PRESS
A DIVISION OF HAY HOUSE

Balboa Press books may be ordered through booksellers or by contacting:

Balboa Press
A Division of Hay House
1663 Liberty Drive
Bloomington, IN 47403
www.balboapress.com.au
1 (877) 407-4847

Because of the dynamic nature of the Internet, any web addresses or links contained in this book may have changed since publication and may no longer be valid. The views expressed in this work are solely those of the author and do not necessarily reflect the views of the publisher, and the publisher hereby disclaims any responsibility for them.

The author of this book does not dispense medical advice or prescribe the use of any technique as a form of treatment for physical, emotional, or medical problems without the advice of a physician, either directly or indirectly. The intent of the author is only to offer information of a general nature to help you in your quest for emotional and spiritual well-being. In the event you use any of the information in this book for yourself, which is your constitutional right, the author and the publisher assume no responsibility for your actions.

Any people depicted in stock imagery provided by Thinkstock are models, and such images are being used for illustrative purposes only. Certain stock imagery © Thinkstock.

Print information available on the last page.

ISBN: 978-1-5043-0781-9 (sc)
ISBN: 978-1-5043-0782-6 (e)

Balboa Press rev. date: 05/31/2017

To my family and all my Guardian Angels

Contents

PART III: THE LESSONS I'VE LEARNED

Introduction

My life has granted me many amazing experiences, some unbelievable, some extremely painful, and others ecstatically joyous. My greatest wish is to share my journey with others who face challenges in their life and to satisfy my mission to have a farmer's story heard. Sympathy is not my intention, understanding is. My livelihood depends so much on the weather and international markets, both of which can change dramatically over night leaving us to stand alone. No other profession has to deal with the environment farmers have to. We live on a double edged sword, day after day, year after year. More importantly my job relies heavily on the integrity and understanding of those I work with.

I am a very proud, honest, and hardworking individual who enjoys being close to the earth and working with animals. I love the wide open spaces, watching the birds fly and singing to their hearts content, and the animals lying out enjoying the warmth of the sun. The simplest moments of delight make my life magical.

The everyday pressure on farmers is enormous; the bankers, county councils, environmentalists, dairy company, animal welfare, OSH (Occupational Safety & Health), the list goes on. Wherever we turn there is a regulation to answer to and more money to be spent. I suspect it will end when every farmer is bankrupt and has to move off the land. But who will produce the food then? Who will rise before dawn every morning, do a manual days work and still be out there at six in the evening nearly 365 days of the year? Who will do all that work for just a couple of dollars an hour? My guess, no one. Well, this is our lot and for many the only perk is when they sell some animals or a small piece of their land. This is the farmer's plight worldwide, but it doesn't make sense to me because I know we are the most important inhabitants on this earth. We grow food.

We are so important I had to write this book giving an insight of how my family survived and how my passion was shaped, increasing my respect and ability to work side by side with nature. It wasn't easy. I dealt with many different people, some of whom made life extremely difficult. As I travelled through the dark times I experienced extraordinary events that I couldn't explain. I have opened my heart and soul to share my experiences and lessons within these pages. It was meant to be.

Preface

A close friend once told me 'life is so boring. I have lived in the same house, on the same property and shopped in the same town all my life'. My world, in comparison has been anything but boring and the complete opposite. I have lived in New Zealand all my life, but since my birthplace in Carterton, I have resided at 22 different addresses in five different North Island provinces. Each place has influenced my thinking, shaped my opinions and guided me on my pathway to the present day. My interest in the farmland, weather and animals began at an early age and gradually grew into the passion that is responsible for how I farm today.

Gypsy Day is synonymous with the beginning of the New Zealand dairy season, falling on the first day of June each year. The date signals the start of new contracts and the day families all over the country move themselves and their possessions to a different location. A large number of farmers shift from property to property working towards their ultimate goal of land ownership. Initially their work is done for a salary. The next step is to become a contract milker where the salary is paid in relevance to the milk production that's harvested. In both of these circumstances the property owner owns the milking herd and will pay for a majority of the services required to get that production.

Many property owners cannot offer the next step however, where the labourer purchases the milking cows and shares the income from those animals and the cost of production in a 50/50 split with the owner of the property. As New Zealand dairy farms have increased in size, the availability of the 50/50 contract has diminished. Fifteen years ago the average herd size was 180, it is now 350. With each animal worth approximately $1600 in today's market, dairying requires huge bank loans to purchase a herd of cows as well as the necessary machinery to apply for a 50/50 position.

My husband and I entered the agricultural dairy sector with the goal of climbing the ladder the same as our corporate cousins in the towns and cities. Our goal was to work hard, earn top money, buy a herd of cows, pay off the debt, and invest the equity into a property, simple you might say. Then I ask, why are not more of us getting the farm we all dream of when we work so damn hard from dawn to dusk, 24/7, in all weathers to reach that final goal? My story might help to answer that very question.

After many challenging years in the dairy industry I concluded time was running out for us. The race wasn't worth the heartache. We had to face up to our situation and we had to *do what we do best, but do it better*. At that stage Fonterra, New Zealand's largest dairy export and processing company was offering a premium for organic milk. It would increase our income and negate the need to invest in more cows, more labour and bigger machinery. It was not to be a smooth transition but it was something my youngest daughter and I were very passionate about.

There is no single description for organic farming and no two farmers use the same recipe on their properties. When the consumer purchases a *certified organic* item the licence number verifies it as an authentic chemical, drug and GE free product presented with a lot of love and appreciation of the environment. BioGro and AsureQuality are New Zealand's main registration agencies and their annual audits must be strictly adhered to by the organic producer so they can earn their place as a certified supplier. Every input introduced to their system must have a documented historical paper trail and these two agencies are responsible for adding their stamp of approval, identifying the end product as authentic and fully traceable from soil to plate.

I have witnessed two completely different approaches to organic agriculture. There are those who continue to use the same type of conventional inputs but purchase the certified organic version to stay within the rules of certification, while the second approach involves individuals who have been able to see and work outside the square. This can lead those working alongside nature to biodynamics, the ultimate organic realm.

Biodynamics is holistic and spiritual and comes from a deeper understanding that hasn't necessarily been backed by today's science. The concept demands we are all ONE in this universe. It's amazing to think every living thing in this universe is surrounded by an electromagnetic field, positive and negative, pushing and pulling, influencing all our daily decisions. With increased awareness of my surroundings and an amazing new book of lessons to learn I changed the way we managed the land, opening our minds to organic and biodynamic methods. Since including the latter two principles into our farming system my job has become fun and exciting again.

My role as a dairy farmer is harvesting the ultimate food from our cows to feed millions of people. On average the milk collected from our girls has fed 6,000 men, women and children every single day for more than forty years! Besides milk, our girls have also provided many other consumables such as cheese, butter, yogurts, leather, fertilizer, meat and more. They are incredible animals and deserve our every respect.

I've discovered over time it's much easier to work with Mother Nature than against her, after all she has been around a lot longer than we have. I also venture into another world of which I was a sceptic at first but then comfortably received advice and guidance.

Earth offers most to those who listen

PART I

GYPSY DAYS

CHAPTER ONE

Early Memories

My life's destiny began the day I entered this world in Carterton, New Zealand in 1956. Carterton was a small rural community where everyone knew everyone, situated in the lower half of the North Island in the Wairarapa province. At first my parents lived in town where my father was a cabinet maker, but a few years later we moved to a sheep station just east of Carterton to a little area called Ponatahi. My sister and I attended the Ponatahi primary school with a total role of thirty pupils catering from primer one to standard six. I was the only pupil in standard five and then standard six, the year before I went off to Kuranui College in Greytown. It was quite a scary transition going from a small country school to a college with a thousand pupils.

My parents were keen to give the family the same opportunities they had, growing and experiencing rural life. When the sheep farm position came up, Dad put his cabinet making skills on hold and moved us to the countryside. He never gave up his trade as he later made family members and friends some beautiful furniture and used his skills to renovate the houses he and mum lived in.

I got my first taste of living in the country when I was seven years old and the freedom was amazing. I was able to run free up and down the hills, through the dry creek beds and along well-trodden sheep tracks. I loved it, challenging myself to run the sheep tracks as fast as I could and I remember being out there till dusk, investigating culverts and navigating the hillsides.

I took an early interest in Dad's day-to-day chores as a sheep farmer and each morning before I walked to school I had to know what he would be doing on the farm that day. I must have been a real nuisance because I was never satisfied until I got the complete rundown of the day's events. I wanted to make sure I wasn't missing out on anything.

November was the best time of the year when I actually got to catch the lambs and pass them to Dad who put rubber rings on their tails so they would later fall off and help prevent fly infestation. They were so fluffy and soft and their aroma was beautiful, so cuddly and warm. I recall the pens that were erected in each paddock to muster the ewes and lambs in for the event. The bleating of the 200 or so ewes and lambs was horrendously loud! The ewes were calling for their babies and the lambs were calling for their mothers. I was amazed how they all paired up again after the lambs had been drafted out and docked. They all sounded the same to me!

Shearing was also an enjoyable event for me and again, I loved the smell of the wool, the sheep's breathe filling the atmosphere in the shed and the natural grease from their fleeces coating my hands. For me the shearing shed was a homely, comfortable place to be and I loved being there.

On the sheep station I quickly learned the importance of the regional weather patterns and how they can influence the land and animals on a daily basis. In the Wairarapa white droughts were and still are customary for the summer months from December through to March and sometimes longer. One hot dry summer day I saw a white rainbow in the sky. It was an eerie sight with not a cloud in the bright blue sky nor a breath of wind moving the trees. I've never forgotten that day and the many cloudless days that followed, one after the other, the sun beating down on the dry earth, cracking it open with its relentless heat. The green grass became shrivelled dry stalks and the fields a white landscape with the odd dusty grey coloured sheep roaming in search of food. Any cold fronts in the weather forecast would move onto New Zealand from the west but the rain wouldn't come across the Tararua ranges. They provided a natural barrier on the western side of the Wairarapa and I used to see the hills covered in misty rain but nothing would come east to relieve the desert like conditions. It was torture for man and beast watching the rain so close but so far away.

I used to spend many memorable hours cooling off in the primary school's swimming pool with my brother and sister. It was really handy, just a short walk across the road. Another memory of those hot summer days were the tadpoles that blocked up the water inlet in the toilet cistern! This was a real curse as the water in the dam began to evaporate and Dad seemed to always be opening the inlet to clear them out. The dam water was gravity fed down to the house just for the toilet and the tadpoles were using the pipe as their way to freedom from the relentless heat and declining water level.

The winter months were also fairly harsh in Ponatahi. The heavy frosts would freeze the water pipes and I used to smash the ice on the puddles as I walked to school. Occasionally they were complete ice packs inches thick and they would stay there for most of the day so I would break them up again on my way home. I enjoyed watching the TV weather forecast and got really excited when snow was predicted. I remember pestering my folks so badly on such a day that Dad just had to take me with him on the farm! Mum dressed me to the hilt with gloves, woolly hat and a thick coat and I was warned not to complain if I got cold. Dad and I went out on the lambing beat that day as the snowflakes began to fall. It was wonderful! I often tagged along with Dad but this was a first in such bad weather. We walked a short distance before we travelled on the tractor to check the ewes. I don't recall how long we were out there, but I do remember not giving into the cold and not admitting to Dad, I was bloody freezing!

I looked forward to the May school holidays when the tent and camping gear got packed into the station wagon and we drove north for our annual family break. I used to keep a diary of our travels and the most noticeable addition was my description of the landscapes and pastures. We used to travel up via the Desert Road to Lake Taupo in the middle of the North Island. I recorded the green grass around Woodville and then the dry brown tussock as we approached the central plateau and the mountains, Mount Ruapehu, Ngaruahoe and Tongariro. I also noted the grazing animals we went passed, the Jersey, Friesian and beef herds, their size, whether they were big or small herds and if they were eating hay, silage or just grass. I was about ten years old when I kept the diary and I am surprised at my interest of the land even way back then!

On one of our annual family holidays we travelled up as far as Kerikeri, nearly to the top of New Zealand. I was amazed at all the orange orchards, the Tamarillos and the beautiful clear blue ocean where the fish were clearly visible looking down into the water from the pier. My diary records how stiff and compact the grass was in places which I now presume was kikuyu grass. Kikuyu was very foreign to me 50 years ago but it has since spread into many North Island regions and I have since grazed the milking herd on kikuyu pastures on three of the dairy properties we lived on.

On another winter holiday we travelled to Taranaki. I can't recall my impressions of the countryside on our journey except to mention the winding roads that made me terribly car sick. Then all of a sudden there was this incredible mountain, Mount Egmont it was known as then and it seemed to lunge up out of the ocean in the middle of nowhere. Being the month of May there was a small coating of snow on the summit and I remember the region being colder than at home in the Wairarapa. It definitely wasn't beach weather as the southerly wind was coming straight off the mountain. We stayed at the camping ground on the beach at Fitzroy, north of the mountain and despite the cooler weather I loved camping out under the stars. I used to lie with my head outside the tent looking up at the stars. What a beautiful canopy I had above me, all twinkling and shining so brightly. I realised the vast spaciousness above me but at the same time I also felt as if I wasn't alone as each star seemed to shine down on me. It was a very vivid and unforgettable experience that I can still remember so clearly.

Back on the Ponatahi Sheep station the property had its own lime stone quarry. This always fascinated me, seeing small seashells in the hillside in the white/yellowish 'dirt' and how this lime was extracted from the cliff face and spread on to the land above as fertilizer. I was fascinated as to how those shells ended up buried underground in the middle of the Wairarapa, some 100kms from our present day ocean. I remember being told the Wairarapa valley was once a huge river and I'm guessing that was after the land got pushed up from the ocean millions of years ago. As I wrote earlier, stories like that always got my undivided attention.

While living on the sheep station I got my first taste of the fertilizer I would learn to despise, phosphate. The day after the fertilizer truck spread the superphosphate on the paddocks, Dad was rather anxious because the rain forecasted didn't eventuate. This was confusing to me but a few days later I began to understand his frustration. A decent shower of rain was needed to wash the phosphate off and without rain the phosphate was beginning to burn the grass by the fourth day. Huge burn marks were appearing where the fertilizer truck had been. Being eleven years old and watching the grass being burnt left a lasting impression. It didn't make sense. I noticed the limestone didn't damage the pasture so why was Dad applying a fertilizer that burnt everything? I didn't ask any questions back then but I have since learned why phosphate was used and the reason why it was the fertilizer many were advised to use in the 1950s and 60s. I also learned phosphate's the pink stuff stuck on the end of match sticks, no wonder it burned the grass! To my astonishment it was an ingredient used in the bombs during World War II. The story goes that after the war the Americans had a stock pile of phosphate and other chemicals they had to get rid of it. To use the phosphate as fertilizer on the land came about by accident. A portion was spread over some grass and a scientist recorded the pastures extraordinary growth! He was overwhelmed because now it could be sold as an agricultural fertilizer. A number of post war chemicals also went the farmer's way. DDT was one that was advertised as good for our vegetables and great for mothers and babies too. DDT was better known as Agent Orange because of the orange paint on the 44 gallon drums it was stored in. Farmers in New Zealand and other countries were persuaded to spray their pastures with DDT or Agent Orange to kill grass grub and other pests while overseas it was used to stop the mosquito transmitting and spreading malaria. It worked alright but the residual didn't break down. It is still in our soils today, many decades later. No one knows how long it will take to dissipate but scientists do know DDT is the most commonly detected pesticide in mother's breast milk today despite its use being banned many years ago. Phosphates should also be banned because along with the synthetic urea used by many, phosphate is polluting our streams and rivers, killing beneficial fungi and many native fish and plant species and doing just as much damage as DDT.

One of the most beneficial fungi on our planet is the humble mushroom that comes in various shapes and forms. I had the ultimate pleasure of harvesting oyster mushrooms by the buckets loads from the sheep farm in Ponatahi. They would begin to grow just after the summer droughts were broken by the warm autumn rains. Fresh sparkling white field mushrooms would sprout up everywhere in their hundreds and my sister and I used to pick box loads to sell to Turners and Growers in Wellington. We used to sit in the middle of the fairy rings and pluck off the best looking ones. We'd cut the stem neatly and pack the mushroom carefully in the wooden box. It was fantastic pocket money! Unfortunately our harvest wasn't as good if the fertilizer was applied before the mushrooms grew and I now know the phosphate was responsible for killing them off.

My knowledge of the mushroom has increased tenfold since those days. They are a well-respected medicinal food and the microscopic cells called mycelium, the fruit of which are mushrooms, run underground for many miles. It is documented mushrooms could help save the world because they recycle carbon, nitrogen, and breakdown plant and animal debris in the creation of rich new soil. They are extremely beneficial to humans as well because they can help our bodies fight many of our modern day illnesses. So wherever the fungal mushroom resides you can bank on it doing a service to mankind and to Mother Nature no matter what colour, shape or size it is.

I took to the outside life and enjoyed helping Mum and Dad in the garden at an early age. I had my own garden patch at primary school and I got a lot of satisfaction keeping it tidy and pulling the weeds out. One of my jobs around the house was to disbud the roses and cut the long grass under the hedge. They were prickly chores but I didn't mind. The job always looked neat and tidy when I was finished. I recall Mum's surprise one day when I took to the Red Hot Poker plant with the hedge cutters pruning it down to ground level thinking it was just a bunch of stalky grass. Lucky for me the plant never flowered so abundantly as after the unscheduled haircut!

My other interest was music. Those were the days when I couldn't wait to visit my grandparents and sing around the piano with my uncles, aunties and cousins who all played various musical instruments. I used to copy songs I'd heard on the radio on the piano and I wrote my own music pieces too. One song in particular has travelled with me all these years:

Man has ruined this earth, Islands there no more

Mother Nature cries

Time will soon run out

Verse 1: Humans came and destroyed themselves, Earth stood to its ground again, smoke and death fill the air so thick, dirt and rust form a mountain.

Verse 2: War has climbed this earth of his, love will sweat its tears no more, life is all but emptiness, death is blessed with joy.

Verse 3: Trees they used to grow so green, shredded figures now there stand, flowers used to bloom the earth, empty lifeless land.

Obviously the earth's future was a significant worry to me all those years ago. I remember there was even a song writing contest at college in which many students expressed their concerns for the future. That was back in 1971.

After all this time my birth place in the Wairarapa still holds my heart. Each time I visit, a special warmth comes over me when I exit the Manawatu Gorge, through the ranges and drive south towards Carterton. It's an extraordinary feeling of comfort and familiarity which seems strange because I only lived there till I was fourteen. Since then I have lived in many other fantastic places and experienced some wonderful milestones but I've never felt the same emotional pull to any other area.

The Wairarapa gave me a truck load of impressionable

memories and one I distinctly remember but would rather forget was when an older college pupil asked me what I wanted to be when I left school. I replied, 'I want to be a farmer'. I simply got laughed at. 'How could you possibly be one? You've never lambed a ewe and got your hands inside to pull the lamb out, not like I do. You will never be a farmer!' I didn't reply, but at the time I was thinking, 'I suppose she's right. I don't do that sort of thing'. I felt as if the rug had been pulled out from underneath my feet, I was absolutely gutted.

The bottom of the North Island was where the interaction between land and animals was first introduced to me and became the love of my life. I enjoyed helping my maternal grandfather and uncle in the old walk through cowsheds during the school holidays, where I could physically touch the cows while they were being milked and I loved coaxing the cows into their *stand-alone* stalls so my uncle could tie their hind leg with the rope and put the milking cups on. I remember being in the wrong place at the wrong time too and being told to 'bugger off back to the house'! I was trying to help Grandad at the time but he must have been having a bad day.

We lived in Ponatahi just east of Carterton for seven years and I was fourteen when I experienced my very first Gypsy Day. Mum and Dad were looking for a change and decided to move from the Wairarapa to Taranaki to work on a dairy farm. Working with the land and animals was in the blood. On both sides of the family, I had grandparents and several aunties, uncles and cousins who were and still are farmers and caretakers of the land. My father's father was also involved with the manufacturing of the end product as manager of the Parkvale Cheese factory until its closure in the 1960s. I can still see my paternal grandfather in his big white hat and white overalls walking into the house and I remember the smell of the whey on his clothes which was unforgettably potent! Each individual farmer used to transport their own milk to the factory in cans on the back of their trucks or horse and cart in those days. Another image is Grandad testing the milk in the large open rectangular vats before he made the cheese. Dad can recall being put into the empty vats to play while his father worked in the factory, the vat being used as a playpen with no way of escaping. Nowadays the milk doesn't see daylight from the time it leaves the cow's udder till it becomes a block of cheddar cheese!

My first Gypsy Day experience arrived June 1st 1970 and with all the household items loaded in the furniture van, Mum, Dad and we three kids piled into the Holden Station Wagon and headed north. We said goodbye to Ponatahi really early, around 4am because we wanted to get to our destination in Taranaki before the furniture truck. It was a memorable trip for me, I was car sick, again! Those damn windy roads. The whole trip took a gruelling seven hours. It was a great relief to finally reach our destination and get out of the car.

My folks hadn't seen our new home before that day and believe me, none of us want to see one like it ever again! The section around the red weather board house was covered in long grass and we had to navigate our way through it to reach the back door. The door was difficult to open as it was catching on the floor boards. Dad walked in first with us all close behind. This can't be the place. Dad must have the wrong address. We walked into a wash house and there was a toilet to the right. I looked in at the loo. The floor was covered with old toilet paper rolls, a heap of dead leaves and the toilet itself was indescribable! I forced the door shut.

The next doorway took us into the kitchen. Mum was already in tears. This was worse than a pigsty! The kitchen floor was caked in brown dirt. There was old mouldy food everywhere. The electric stove was black, not white and there was an old coal stove in the wall covered in rubbish. Dad opened the adjoining room. Oh my God, this was where the chooks had been kept! The room wreaked of chook shit or something dead! Needless to say the door was slammed shut. This just wasn't happening, it was a ghastly nightmare!

The other adjoining room to the kitchen was the dining/sitting room. Not much going on there except the brick fireplace was full of junk, tin cans, beer bottles and partially burnt newspapers. By this time we were all completely lost for words, we were dumbfounded! Surely this can't be right. Dad couldn't contact the owner because he lived fifty miles away and the phone was disconnected, we were on our own. No friends or neighbours to turn to and no mobile phones back then! Worst of all we had a truck load of furniture less than three hours behind us and it had to be unloaded into this hovel? Mum's tears and Dad's disgust was felt by all of us. The bathroom was worse still, with a huge hole in the floor boards under the bath where rats had entered leaving their droppings everywhere. The bath and wash basin, long past their retirement age, were absolutely black with filth. We had no choice. We had to start scrubbing and sweeping so we could at least empty the furniture truck, get a bed to sleep in and prepare something to eat, just bread and butter looked good at that stage but where the hell do we start?

The Station Wagon was full of the cleaning equipment Mum and Dad had used to leave the Ponatahi house spotless on our departure, so they got it out of the car and started in the kitchen, removing the old food from the bench and cupboards, heaving it out through the large slide up window. My younger sister and I started sweeping the board floors in the three bedrooms and the living room with a small brush and shovel. That was all we had, all the big brooms were in the furniture truck. There were no carpets in the rooms, only polished floor boards covered in dirt and rubbish. I guess Mum had to clean the toilet. Thankfully none of us were in a hurry to use it. We left Mum and Dad to themselves and did what we could at the other end of the house. In other words, they didn't need any of our moans and groans to make matters worse.

The fourth bedroom, (the chicken shed) just off the kitchen was an enormous task in the days that followed, scraping the manure off the floor and walls. It's probably difficult for an outsider to imagine the state of the house we were asked to live in, but it was even harder for us to make it liveable in those first few hours with the race against time and the arrival of our furniture. We had to keep going and keep our spirits up. At the end of the day after all our wonderful mother had been through, she inspired us all with her words 'you can always make a house a home'. Her philosophy has stayed with me to this day and I have applied it to every home I've lived in.

Three weeks after our arrival in Taranaki, a home is exactly what my parents had accomplished. The house had been scrubbed from top to bottom and some of our belongings had been used to cover the cracks in the floor boards and holes in the walls. The grass around the section had been trimmed and mown with the old flower beds resurrected. The old house became our home and our pride was restored once more. It has always baffled me as to the conditions the owner expected us to live under. No one should be expected to go through what my parents did, but I was to encounter a similar situation a few years later.

As of 1st June that year Mum and Dad became dairy farmers in Inglewood. Being the eldest, my challenge was to fit in at the local High School while my younger brother and sister were enrolled at Kaimata Primary School. After school I really enjoyed helping with the calves and on occasion milking the cows. The biggest challenge was finding new friends and navigating around a new district.

At High School I studied Biology, Science, History, Maths, English and French. My favourites by far were English and Biology which have played a big part in my later life. I didn't realise how my future interests would be based on those two subjects. I just wish I hadn't thrown away all my fifth and sixth form Biology notes. I kept them for years, through shift after shift until finally I decided they were taking up too much space and they had to go. Isn't it always the way, the minute you ditch something you regret it and want it back!

In Biology I recall studying plants and animals and all the associated fungi, algae, bacteria and smaller realms with which we live side-by-side. It was all very well studying them in books but it wasn't until my late 50s that I came to realise how delicate our ecosystem really is and the huge role that's played by everyone including all the bugs and bacteria. I actually forgot about all that *life* in the soil. At High School I didn't see the relevance in learning about it, I was just copying stuff off the blackboard and storing information in my head to pass the exams. It wasn't until I actually became part of the real world, dealing with reality that I began to learn things for a reason.

CHAPTER TWO

One & One Make Six

Over the following years I finished four years of high school, being accredited with a University Entrance, got married, had four children and began to work with hubby in the dairy industry. I remember in the mid-1980s David Lange, Prime Minister of New Zealand at the time insisting farmers run their properties as a business and no longer treat it as a lifestyle or a hobby. Several rules were changed to mould us all into the new business role. It was about that time when subsidies for the industry were abolished as well and the Government wasn't going to prop up our agricultural sector any more. The subsidy removal was a benefit to New Zealand but it was about this time farmers lost a personal touch with the land and were forced to focus more on profit and extending their businesses for capital gain.

My first experience of moving house with a husband and young baby came on Gypsy Day June 1st 1974. We shifted to Otorohanga in the King Country to work as contract milkers on a 170 cow dairy farm. We packed all the furniture into the removal van, chucked the cat into the car and headed north from Taranaki. Our new home was a comfortable three bedroom cottage and it was clean and tidy to move into. We had a great season that year, exceeding the property's best milk production. It wasn't necessary to feed supplements such as palm kernel, maize silage or meal because the grass pastures were ample and thinking back the animal health was incredible. I recall the beautiful heifer calves we reared that year, all taught to drink milk from their own individual buckets. We didn't worry about target live weights as is the recommendation nowadays, because the calves were fed for at least sixteen weeks on whole milk and had pasture available to them from day one. Veterinarians and farm supervisors support the view calves can be weaned at twelve weeks of age as long as they weigh around 90-100 kilograms for the Friesian heifers and 85-95kgs for Jerseys.

We were on cloud nine that season and so decided to milk more cows and look for another job with a bigger herd.

Gypsy Day 1975 we packed all of our belongings into a couple of cars and trailers and moved to the Okoroire district in South Waikato to milk 240 cows. Our new contract stated we had to employ a third person to help which consequently made things very difficult. Looking back we were too inexperienced to cope with the position as well as employ a worker who was older than us. It was an experience I would rather forget. I guess the day we moved in was the writing on the wall. The house was nearly as bad as the one I had to help my folks clean up a few years earlier in Taranaki. The kitchen had rotten food in the cupboards, the stove was blackened on top and the oven full of old smelly breadcrumbs. I had to literally shovel the crumbs out of the oven. I recall not being able to see the pattern of the lino on the floor in the kitchen/dining area until I had time to scrub it a week later. The rest of the house had rubbish in every room and looked as if it hadn't been vacuumed for months. It was disgusting. How do people live like that, and the former occupants had a young baby crawling around the floor? The house was surrounded by long grass that hadn't been mown for months and the gardens were non-existent.

The house and section were a challenge, as was the farm. It was a very wet property with contoured humps and hollows in the paddocks to help with drainage. It was quite a sticky soil, unlike Otorohanga where the property was a gently rolling contour with free draining. In both places we grew all our own vegetables and I planted a lot of flowers and shrubs around the house using cuttings taken from Mum's garden in Taranaki. It was very satisfying. In those early days I used to have heaps of young plants growing in pots ready for transplanting. Forty years on I now fill the garden beds with plants and herbs that are edible as well as colourful and I've stopped carting the numerous pots of young cuttings from place to place.

Unfortunately the Okoroire milking contract was not working for us nor the owner so we packed all our belongings on to a couple of car trailers and moved to another job. This all happened in January, the middle of the dairy season. We were lucky to find a fantastic position working for an awesome couple on the outskirts of Cambridge. It was a herd of 130 cows on a perfectly set up dairy farm. All flat contour with huge barberry hedges for shelter in every paddock.

At this stage I was two months pregnant with our second child who was born a month before our next unscheduled move.

We experienced an amazing dairy season in Cambridge with plenty of rainfall during the summer months of January and February. In fact there was so much rain, the hay crop had to be heaped up and set fire to. There just wasn't enough dry weather to be able to get the hay baled. There was no summer drought that year. I must confess, I didn't have a lot to do with the day-to-day farming chores in Cambridge, I was pre-occupied with things to do around the house like cooking, sewing and doing the gardens. I often preserved fruit and froze the surplus vegetables and there was always fresh meat and milk made available from the farm. Our second son was born in early July. Both of us caught a cold in the maternity home and weren't allowed to go home until ten days later. It was a very long ten days and as I left I was advised of what not to eat in case the baby got an upset stomach from my breast milk.

I had breast fed our eldest son for four months and when I wanted to stop I was advised by Plunket (NZ child care association) to slowly wean him onto boiled cow's milk. I wasn't permitted to give him raw cow's milk until he was at least a year old. The boiled milk was also watered down before I gave it to him. In more recent times, just seven years ago, our daughter gave her son a wonderful milk formula made up with raw organic cow's milk. He started drinking it as a baby and the recipe from The Weston.A.Price Foundation was as close to mother's breast milk as it could possibly be. I'm not sure how the child care organisations of today would respond, but our grandson has grown perfectly well. I suspect they would only allow the powered milk formulas for baby because of the potentially harmful bacteria in unpasteurised milk.

In 1976 I didn't question the advice given by doctors and nurses. I never felt I had to. They were the experts, not me. Now, forty years later I have become older and wiser and I question everything. We have entered the twenty first century and like others I have become aware of how mainstream science is failing us, and how easily it can be manipulated to benefit big industry. The medical professionals are excellent as far as their tuition has taught them but very few of them, if any have knowledge of the basic nutrition or vitamins and minerals required for our optimal health. Our medical doctors study for many years and in that time there is only one book with one chapter on nutrition. Their peers were not taught the benefits of nutrient dense food and it's not in their medical text books so it can't be important. The agricultural industry as well as the medical seem to have followed a similar path. It's all about making money. The faithful family doctor of my grandparents' day that used to make house calls has long since disappeared. It's scary to see reports from America showing how medical professionals can be *bought* by Drug Companies who offer fantastic holidays and money incentives as rewards when the doctors sell their products.

During my second pregnancy I was prescribed iron and fluoride tablets by my GP. I thought nothing of it. I simply obeyed and followed her instructions. I recall being told the fluoride was to strengthen mother and baby's teeth. Thinking back I find it strange that I was prescribed fluoride and iron in my second pregnancy and none of the other three. Were we the guinea pigs of the year or was there an incentive to give them out? Not sure what the fluoride has or hasn't done to our health but I do know I would flush it down the toilet if it was me today. It is a poison and shouldn't be anywhere near a growing foetus. In addition to increasing the risk of a stillbirth, fluoride also calcifies the pineal gland and hardens arteries. I discovered many statements suggesting the harmful effects of fluoride including those from Professor Dr A.K. Susheela of India and Dr Edward Group of the Global Healing Centre. I thank God my son and I both came through the *experiment* okay.

After breast feeding our second son for 3 months he was fed powdered milk formulae. This was when the first baby formulas were developed. These would also have had some fluoride in them. I can only imagine what could have eventuated if I had not insisted on eating our own home grown vegetables and meat and adding wholesome raw milk to our meals. Good nutrition has shown to be a benefit against fluoride damage.

My babies ranged from 6lb 6oz to 7lb 8oz and these weights were considered a good average. A few mothers gave birth to 8lb babies but hardly ever beyond 9lb. In this century less fall into that weight zone and more are born 9-10lb. What is happening? We can't look towards better nutrition because it isn't true. Our food is less nutritious now than ever before.

Coughs and colds, bumps and bruises all came and went in every farming family. The boys got on the tractor with us and out with the animals in the fresh air. They were bundled up and taken with us everywhere in all weather. OSH (Occupational Health & Safety) wasn't heard of then but that didn't mean to say we ever put ourselves or our children in harm's way. The children grew up aware of dangerous situations and learned how to sit on the work vehicles safely and how to walk around animals.

Our contract milking position in Cambridge came to an end on the 1st August, when the property changed ownership due to the owners marriage break up. The property was turned into a horse stud so we were the last to milk there.

Our unscheduled Gypsy Day saw us moving to our next dairy farming position just north of Morrinsville. The position required two full time milkers, hubby and myself but the house was 700 metres from the dairy shed and I had to take our new baby and the older son to the shed every milking. I still don't know how we did it! I was breast feeding at the time and I remember the afternoon milkings when I had to sit in the car to feed baby and change his nappies with the machines going in the background. In the mornings I would speedily give baby my breast milk, tuck him down and leave both of them asleep in the house. At the end of milking I would return to find them both still sleeping, thank God, but as they grew older, this wasn't always the case. On occasion the older one, aged four would be keeping the baby happy by throwing toys into his cot until I got there. As other families would know, the oldest siblings had to grow up pretty quick as they were relied on to entertain and keep the younger children occupied while Mum was working out on the farm. At three months of age I had to introduce powered milk formulae to my son. I was getting extremely tired and my breast milk was declining. Our second baby was born during the busiest time so it was a huge challenge to have his feeding schedule work around me having to go to the milking shed. The bottled milk made it easier for me to be in two places at once and I relied on the older one to help feed his little brother when I couldn't.

The Morrinsville district is one of the first areas in the Waikato to dry up in the summer. One Christmas Day I remember being outside the house begging for the black thunder clouds on the horizon, to blow our way. They were all around us and we desperately wanted rain. The lightning put on a spectacular show against the black clouds and the storm was so close you could smell it, but not a drop fell where we were.

We lived and worked on the Morrinsville property for three seasons. It was a relatively flat property with the odd gentle gully offering natural drainage. The summers were long, hot and dry but the winter months, cold and wet with the soil easily turning to a bog if the herd wasn't taken off the pasture and rested on the concrete stand-off pad.

Our third baby, and eldest daughter was born in April at the Morrinsville Maternity Home. I will never ever forget the love and pure joy I felt the moment I saw our baby girl for the first time. No matter what gender the baby was going to be the love would have been the same, but to experience a little girl after having two boys was something special. Hubby had to go home to milk, so he missed all my hard work.

With three children now, my days were rather busy. A few weeks after our baby girl's arrival, hubby's cousin offered us a job on his newly purchased dairyland on the boundary of Te Awamutu. I recall not being too excited because I was enjoying the Morrinsville area, the oldest had just started school and there was a young baby to think of as well. I was persuaded it would be a positive move so the furniture removal truck was ordered and sent off with our belongings to Spirit Road, on Gypsy Day 1979.

Te Awamutu was to remain our home town for the next 23 years. The youngest of our four children, our second baby daughter was born there and all four of them attended the Te Awamutu Primary school, the Intermediate and then the Te Awamutu College. It was great to feel grounded and live in familiar surroundings, making lifelong friends. The years went by too fast, probably because I was helping on the farm, milking twice a day as well as doing all I could to support the family with their sports and interests.

Hubby's cousin was a livestock dealer, buying and selling dairy cows as a profession. The plan was for us to milk the base herd of 120 cows as contract milkers, while he was buying and selling the other fifty. It was a very demanding position and heart wrenching when the favourites were sold off. Nevertheless, we learned a lot. We were taught to pick the good cows from the bad by looking at the animal's structural make up. Looking at their pin bones, their shoulder width, how high the udder was held between the hind legs, the depth of the gut, and then their behavioural patterns, the list went on! Basically we learned to place a good animal into a box, you might say. I know that sounds ridiculous, but as an overall description that is how the old timers knew the best milk producers from the worst. If you look along a cow's strong, straight back, down the hind quarters, under the belly and back up, there is a square, or a slight rectangle. The head should be strong and upright and the chest a good width for a strong heart. We have used these attributes over and over again and they have not put us wrong yet.

The dairy shed on the Spirit Road property was very close to the urban area and I have often wondered if that was the reason the cows were plagued by biting flies during the summer months. Every season from December 12th until March 5th we had terrible fly problems especially during the afternoon (4pm) milkings. The shed was a twelve aside herringbone and we could not have both sides full of cows at the same time. We had to milk one row at a time and even then it was a challenge to stop the cows kicking cups off because the biting flies were giving them hell! The water from the high pressure hose did dull the problem for a short time but the flies would come back as the next row of cows came in. The flies really took the fun and enjoyment out of milking and the hot temperatures made it unbearable at times.

Most days I took home a billy full of milk for the family to drink, put on our breakfast cereals and use in my baking. We had become a family of six by then, with my life's wish of bringing two sons and two daughters into this world, a dream come true.

One day a farmer's wife asked me if we drank milk from our own cows. I said, 'yes we do' and to my astonishment she replied 'Yuk! With all the shit that goes in it, I'd never drink the stuff from our place!' Wow, this was a farmer's wife. I still find her comment unbelievable. Our pride demands, *if the milk going to the dairy factory isn't good enough for us to drink, then we shouldn't expect others to drink it.* I stand by that principle to this day.

On Gypsy Day 1986 we were at last moving up the ladder after working with hubby's cousin for several years as contract milkers. That year we moved west of Te Awamutu to Peat Road, purchased 180 jersey cows, and entered into a three year 50/50 herd owning contract. As described earlier, this type of contract is when the property is owned by the landowner and the labourer owns the cows with the income from the cows' milk production split 50/50. Basically, the expenses involved to achieve that production falls on the sharemilker while the owner of the property has to ensure the fertilizer is applied and the machinery in the dairy shed is in working order.

We were really excited to begin our first sharemilking contract in 1986. It had been a long time coming. Saving enough money to get to that stage wasn't easy and we now had four young children requiring financial support for their schooling and sports. Every school day the children had to get themselves up and out of bed, get their own breakfast and walk 800 metres to the bus stop. The breakfast table and the school lunches were always organised the previous evening and if I wanted to communicate with the children before they went to the bus, I would leave a note on the breakfast table the following morning. Occasionally it worked the other way and their note to us would be a reminder of a school event or some money to pay for books. They very rarely missed the bus and were rarely sick. We had four of the most amazing children!

The Peat Road farm was situated on a black peat soil. In this area you thought twice about burning fires in the dry periods. If the fire spread to the rotting stumps and old trees buried beneath the surface it could take days or even weeks to extinguish, and could easily spread for some distance. In the winter the land rocked as the cows walked from paddock to paddock because of the high underground water level. If the ground was cultivated too deep, the top crust would be disturbed, effectively bogging the tractors. Heavy rainfall wasn't a problem because it could rapidly filter through the crust. The beautiful black peat soil was sort after by horticulturists to use in their potting mixes. The property was completely flat with several drains and very few trees to provide shade for the stock during the hot summers.

In our first season we exceeded the farm's record production, but the second year dealt us a very dry summer and the cows were dried off early. The third season was our biggest challenge. The property changed ownership and the new owner from Holland introduced us to our first farm advisor to supervise us and our management of the property. Her first observation was the production had to increase so we were instructed to apply $8000 worth of urea. That was a cost to us of $4000 as well as $4000 to the owner.

This was our first experience with urea. For those not familiar with it, urea is synthetically derived from natural gas and spread on the pasture to force the grass to grow faster. It is not cow's urine, as I discovered many people think. This was duly applied but we had little to no growth response from doing so. Three weeks later our lady farm advisor returned insisting we apply the same again. We said no. That was in June, mid-winter, and the soil was far too cold for any grass to grow and we had already spent $4k for nothing. This kind of advice soured our relationship with the new owner. Firstly the inexperience of the supervisor and then the urea that was being pushed as the new wonder drug to force grass growth that wasn't going to happen without the right ground conditions. Hubby's cousin had recently returned from a trip to Europe. It was in the Netherlands he witnessed the use of urea and how they were in dire straits because it had filtered down into their underground water reserves causing contamination. His advice was 'do not go down that track and stay away from urea', but we were only witnessing the beginning of New Zealand's obsession with synthetic nitrogen and our young farm advisor, fresh out of university was all for the idea. Our experience accounted for nothing and the introduction of advisors to the arena simply made a mockery of us and other life time farmers.

Gypsy Day 1989 we used car trailers to shift all our belongings a few miles away to another 50/50 position on Swamp Road, south east of Te Awamutu. It was a three year contract. The cows were all loaded on to trucks and shifted with us that same day. The day of our arrival, the property owners went on a four week holiday! That was nice. We didn't know where anything was, the electric fence controls, the water pump, the paddocks with grass to graze, even the initial boundaries! We were completely in the dark as the cows came off the trucks. We opened the gate to the nearest paddock and had to sort the stock the following day.

The owners had been remodelling the dairy shed at the time and it looked like they got tired and just walked out. We couldn't believe it. It was a proper shambles! It took us several days to clean up the cowshed yard and cart away the broken concrete and old timber. The house was rather small but had been left clean by the previous occupants who were renting it. We were the first workers to be employed.

The first season was a challenge, discovering the good paddocks from the slower growing ones, finding why electric fences weren't working, watching the grass being pulled out of the ground as it was grazed and the poor drainage areas. We had it all. This property had a river running through the middle of it which was inclined to flood when there was heavy rain up in the hills. We had to make sure the herd was on the home side of the river so we could get them to the milking shed. One day we were travelling back from getting the groceries and we noticed the river had risen dramatically. The race was on to get home and bring the girls closer to the shed before the lanes became impassable. I remember thinking, 'hell, we don't even get a break to shop for groceries!'

In our first season our girls produced well, smashing the previous record and we felt really proud. The second season began and the owner introduced us to *his* farm advisor. Here we go again but we wondered what we had done wrong and why was an advisor needed? This time he was an older guy and had been a farmer so we felt comfortable with the prospect of working with someone who would be on the same page as us. That couldn't have been further from the truth. After his first walk round the property his written report followed. 'The gates had been left open after the fertilizer truck' as well as 'The cows should be eating x amount of grass and producing x amount of milkfat' which we'd heard before. One thing that really got our backs up was we had to cut down and remove all the random barberry bushes in the fence lines. There were quite a few and those bushes had been growing in the fence lines for years, but the advisor was making it our job to remove them. We were also instructed to manually pull the stinking Mayweed from a huge area where rubbish had been dumped and then buried with truckloads of dirt and gravel taken from the roadsides. Surely that wasn't in the contract? I checked our 50/50 agreement and it read 'the sharemilker is to improve and maintain the property'. We did as we were instructed and then halfway through that season we found out why the farm advisor was introduced, the property was put up for sale! We had been used to smarten up the place and get it ready to put on the market.

Both of the advisors we dealt with wanted control over the bulls we selected to mate our cows with. They intended to increase the herds breeding worth and get more production from their future progeny. While that was our vision as well our breeding plans weren't acceptable to them.

We were at the stage we couldn't see any benefit in being a sharemilker. The landowners were allowed to bring in a third party to run the farm, so why did they need us? We had been farming all our lives, and surely we had been employed on our merits, not someone else's ability to tell us how to do things. We cared for our animals and year after year supplied grade free milk to the dairy factory. We also got into the top five percent of herds in New Zealand without advisory supervision and our girls achieved a fabulous 420kgs MS (milk solids) per cow in 2001, a few years later. It wasn't just our stock's breeding that accomplished the amazing milk production, it was dependent on the number of days the cows were able to milk due to the fantastic weather conditions we had that season.

Mother Nature has full control.

Somehow word got around we were looking for a new position and one day a farmer from around the corner called in and offered us the lease of his property. We negotiated a two page lease agreement to be renewed annually. We decided to sell our herd of cows and purchase the one on the property which had many years of nominated breeding behind it. In fact it was one of the first to be involved in Livestock Improvement's artificial breeding (AB) programme in NZ. Leasing the property meant we had full control of the day-to-day events and there would be no farm advisor involved. Two conditions of the lease agreement included sowing Tall Fescue after the maize silage crop was harvested and the owner was to order and pay for all the fertilizer. I remember him saying 'the first thing that happens when the milk pay out drops, is no fertilizer goes on'. Made perfect sense to us. So all we had to do was pay the monthly lease fee and look after the property as if it was our own.

CHAPTER THREE

Turning Point

Gypsy Day 1991 saw us packing our household items into the car and trailer and repeating the episode several times until we completed the move to the new lease position. It was only a couple of kilometres around the road and we were moving into a lovely old villa, at least 100 years old. There were four bedrooms in the house plus a sleep out with two more rooms. We made this our home for seven happy years. The lease agreement gave us more freedom to rear extra heifer calves and it didn't matter if the milking herd was short of numbers, the complete opposite to how the sharemilking contracts were implemented. In the latter, the owner was entitled to half the value of any surplus animals reared and the herd numbers weren't allowed to go below the minimum written in the agreement.

A lot happened in that seven years on Short Road. In that time the children became more independent and my position changed from *farm hand* to *farm manager*. I was handed more responsibility as hubby left the milking shed and began working as a livestock agent. It was great because now we had an extra income coming into the household. Our second son left school to work with me and if hubby's extra income could pay for his wages, plus a little bit extra, we were on a win. The two girls also played a huge role as extra hands in the dairy shed and helping to rear the calves.

The Short Road property was a turning point for me in so many ways. I began to take a lot more notice of the cows, the soil, and our own personal health but it didn't happen overnight, it was a rather slow process. In fact it would take me another fifteen years to completely stop *following the crowd* and begin thinking for myself.

The 1991-92 dairy season was reasonably kind to us and our 185 cows produced 64,000 milk solids. We planted four hectares of maize that was later harvested and stored in a dirt pit and we used urea to speed up the grass growth in the spring. The tall fescue germinated well and we applied urea in its early growth stage, a practice I have come to frown at. Hay and grass silage were also harvested from the property. There were only small square hay bales made back then and we often got the rugby club members to pick them up and stack them in the barn for their fundraising. The grass silage was cut in early October, swept into a second dirt pit and covered so it was available to feed out in the dry summer months. All-in-all we purchased a lot more heavy machinery to cater for the number of different supplements we had to feed to the cows.

The fertilizer was supplied every year as per our agreement, 15% potash in the spring and 30% potash in the autumn and it was our job to spread it. I recall one spring our son was spreading the fertilizer with the tractor and Max, the Border collie went with him. Max loved to bark and chase the wheels of the tractor, driving us all crazy at times. I was in the milking shed that morning when the tractor returned for another load and our son came into the shed. He came to tell me he had just run over the dog. 'Oh no, not Max, is he badly hurt?' 'Nah……. he's dead and I've put him down the offal hole'. 'What?' I didn't expect that for an answer! I just got the news of the dog being dead and then to hear he was already ten feet under, I didn't know whether to laugh or cry. I thought Max deserved better than that though, a proper burial at least. He was a lovely animal and the girls used to have a lot of fun dressing him up and making him jump through hoops like at a dog trial. He was a family pet rather than a proper cattle dog, in fact if a cow looked at him he used to run as fast as he could in the opposite direction! I later lifted the concrete lid of the offal hole and blessed our poor Max with some flowers and kind words.

The owner's fertilizer mix went on the property every year and our goal was to get production so we invested heavily in 'additives' as well as the machinery. We manually drenched the cows during each milking (twice a day) for at least eight months of the ten month milking season. It was Causmag (calcium/magnesium supplement) to prevent milk fever and grass staggers in the spring and then it was straight into drenching with bloat oil to prevent the cows dying from bloat. If we were lucky, we would get a few weeks break in November/December before we were back into it, this time drenching with zinc to prevent facial eczema. Despite our interventions we were surprised at still having cows with milk fever and facial eczema. The zinc drench depleted the animal's copper, detrimental to their proper mineral balance so we had to give a copper injection every year after the eczema season. The property was a huge learning curve for us and dollar after dollar was going into keeping our stock alive while we attempted to produce as much milk as possible.

We reared the replacement heifers on whole milk and at eight weeks of age introduced them to meal pellets. When they became used to the supplement we weaned them off their milk. I remember how the merchants used to promote feeding meal to young stock in all the agricultural magazines. There was this picture comparing the internal stomach of a calf who was fed meal to one who didn't. One calf's stomach was what I would call of *normal size* and the other was enlarged. The associated slogan read, 'Feeding meal encourages the animal's stomach to grow into a larger healthier organ'. I never used to think twice about such supposedly factual statements but I do now. I suggest and I suspect the enlarged stomach was anything but normal. I suggest it was the irritation created by the introduction of a foreign food that caused the gut to swell. Bovines are equipped to eat and digest foliage, such as grass, hay and silage. Birds have a crop and only they can digest grains and seeds, the gut of the cows and their offspring cannot.

We always reared strong, healthy replacement stock on milk, grass and hay. It wasn't until the idea of putting more milk in the vat, by using the cheaper alternative of bagged meal, that merchants enticed farmers away from traditional practices. Like others we were encouraged to introduce many products to our annual calendar that had been proven to increase milk production, such as rumensin (milk booster), eprinex (de-worming pour on), and urea to name a few. We religiously de-wormed the young stock every six weeks for the first year of the animal's life as recommended by the product's manufacturers whether we saw a worm burden or not.

With the drenching, the urea and supplementary feed added to our system we thought we had everything covered and we were on to a winning combination. But we didn't allow for a decrease in our milk payments due to the international markets nor the dry summer that reduced milk production. Most importantly, we never considered the high cost to our health.

Four years after moving to the Short Road property, we experienced some scary family health issues. For me personally, I had been checking the moles on my back and one day noticed an odd one just below my shoulder blade. I had the doctor check it out. He took one look and said 'right, I'll see you in here again tomorrow and we'll take it off' I was in shock, wow, that quick! The mole was sent away to be tested and to my astonishment it was a malignant melanoma. I wanted to know why me and would there be others? No one could give me a conclusive answer. I had regular checks for the next three years by which time I was given the all clear.

That same year, our youngest contracted hepatitis ten days after being immunised for it at her school. I noticed her skin had a yellow tinge and she was very lethargic. The doctor couldn't explain how she had caught it but in my opinion the vaccine initiated the disease. There was no other explanation.

The doctor was very worried about her liver but said there wasn't a lot we could do, just rest and plenty of fluids. As a mother I didn't appreciate that advice, there must be something? I went directly to the Health Shop where I was advised to give my daughter milk thistle tablets to support and cleanse the liver. Meanwhile the doctor had instructed us to come back in three weeks for a check-up. It was a long three weeks with our little girl missing out on school and being stuck at home. We returned to the surgery for the check-up and the doctor nearly fell over backwards. Our daughter's liver had made a complete recovery! What a relief! I was over the moon to say the least. I told him about the milk thistle and he seemed baffled as to how it could have worked and in such a short time. The milk thistle was the only additive she was taking so there was nothing else to give credit to and her liver was definitely in bad shape before she started taking the tablets.

About six months after the bout of hepatitis our daughter had tonsillitis. At the time her immunity was very low. She had never had trouble with her tonsils in the past. The doctor put her on a course of antibiotics, then another one, then another one! After the fourth visit to the doctor I decided this could not go on. She was finally placed on the waiting list to have her tonsils removed but there was no certainty of when. Meantime I was not going to have my daughter scoffing antibiotics until that time came around.

I again went to the Health Shop to see if there was anything she could take. It was suggested she take Immune Support, a liquid herbal mixture. It was in fluid form and she started straight away with 5mls per day. Our daughter finished taking antibiotics and switched to drinking the Immune Support, daily for the next nine months. At that time her tonsils were removed. The surgeon couldn't believe how *rotten* and shrivelled up her tonsils were. He hadn't seen anything like it before and he was astonished how they hadn't been affecting her health at all.

In my mind that was strike two against the medical fraternity. I had had to do my own research in both cases and I hate to think what our daughter's health would have been like if she'd taken antibiotics for all that time she was on the waiting list to have her tonsils removed.

Our eldest daughter also had a health challenge that year. One day she came home from college and said 'hey look, I've got all these bruises on me'. I asked if she could remember being hit with anything or had bumped into something, to which she replied, 'no, I can't remember anything'. The following morning I took her to the doctor for a check-up. Some blood samples were taken for testing and the doctor didn't seem too concerned after giving her a complete examination. I dropped our daughter off at her friend's house because the girls had planned an afternoon shopping in Hamilton and I drove back home. At approximately three o'clock that afternoon, the phone rang and it was the doctor's nurse. 'Your daughter, is she there? You are to take her to ED (Emergency Department) at the Waikato hospital as quick as you can. Her white blood cell count is extremely low and if she gets bumped or hurt in the slightest, she could bleed to death!' I was in utter shock! What the hell was going on? To make matters worse I couldn't contact my daughter, she was shopping somewhere in Hamilton!

I took a deep breath, several in fact, and convinced myself, 'she'll be fine, she'll be okay.' The clock moved so damn slow that afternoon. Eventually she returned to hear the news. 'They want you in hospital'. 'But I'm not sick!' she insisted. 'I know, but something must be wrong'. I was as bewildered as she was.

We arrived at Waikato Hospital ED around 6.30pm, filled in the forms and yes, they had been expecting us. We sat and watched the sick children come and go as well as some elderly folk in quite stressful states. We read magazines, chatted, and watched a bit of TV. Time past. At 12.30 am the duty doctor at last turned his attention our way. Good, we are on the move, I thought. He said 'You can go home now providing you are very careful about any little knock or bump. Come back tomorrow, no, I mean today around 9am and we will find a bed for you'. I was furious, and the daughter more so! There was no explanation given as to her illness, or whatever it was, and we had been sitting there for six and a half hours! We appreciated the fact the visibly ill babies and others had been given pride of place, but surely we deserved more of an explanation than the one we received.

We arrived home to our beds around 1.30am only to return to the hospital a few hours later. Our daughter had more tests and was placed in a ward with visibly ill patients. She felt guilty, because she didn't feel sick and surely there was a person in more need of a hospital bed than herself?

Four days later she was sent home with a bunch of pills designed to increase her white blood cell count. The doctors didn't know what was going on except to say her body was killing off her white blood cells and suggested it was her spleen causing the problem. Two weeks later the specialists knew the pills weren't working so they tried steroids but with these came weight gain and irritability. Nothing seemed to be working. I took our daughter to an alternative medical practitioner in Tauranga. The solutions prescribed were homeopathic but it was going to take a month to six weeks to work through the prescribed programme. The medical specialists wouldn't have a bar of it and they insisted we didn't wait any longer. They wanted our daughter's spleen removed as soon as possible, because she could not stay on the steroids without them affecting her health and wellbeing. The doctors still didn't know why our daughters spleen was acting like it was, they were completely baffled. To me, their scheduled surgery was taking the easy way out. Within a few weeks of our daughter's surgery I heard of two other healthy teenage girls with the same symptoms and the same result which didn't make any sense to me, it just didn't add up.

In reflection, we should never have given in to the doctor's demands but he was the expert. I believe if the body is born with it, it is there for a reason and the *slash and burn* protocol should be seriously questioned with any alternatives thoroughly investigated before it's carried out. I later learned the function of the spleen is to strengthen our immunity to everyday illnesses and is known as the *mothering organ*, both of which are important reasons not to have it removed. Our daughter's spleen was ultimately removed via keyhole surgery and her blood count returned to normal a short time later. There was never an explanation from the doctors as to what went wrong, just 'cut it out, and she will be able to live a normal life without it'

As I mentioned earlier, we didn't factor our own health into the seven years we lived and worked on the Short Road property. We had never experienced such medical dramas before and we haven't since. I didn't consider the illnesses to be random because of the short time frame around the three of us having medical problems and I didn't seriously make the connection between our wellbeing and the health of the cows and the soil until many years later.

We weren't aware the consequences of poor nutrition were beginning to have an effect on the herd with the number of empty (not in calf) cows increasing annually with a lot of young ones being sent to slaughter because they weren't getting pregnant. We began to introduce

CIDRs into our herd management. These were Controlled Internal Drug Release capsules that were inserted into the cow by the vet. We used a few the first mating season, then more the following year and more the next. We started to look for reasons why the cows weren't getting in calf. We were blaming the bulls not doing their job properly, the consequences of eczema, or maybe the girls were using too much energy to produce milk and not releasing enough fertile eggs for reproduction. All these ideas were only masking the real problem.

The true indication something is wrong can be seen in failed reproduction. This is indicative of our human population as well as the animal kingdom and I only have to look at the rising number of fertility clinics in our towns and cities to see the evidence. Something is wrong.

One of my major concerns centred on the property's soil fertility. The paddocks growing Tall Fescue weren't growing as they should have been, so I made some enquiries through Massey University. After seeing the soil tests, they concluded the soil wasn't fertile enough to support the Fescue. Now all we had to do was convince the owner. Surely he would see something wasn't right. The property had seen a lot of urea applied to the pasture over the years and the sulphate reading in the soil test was between 290 and 400 units where it should have ranged between ten and twenty. The soil scientists at Massey couldn't explain why sulphate was being stored in the soil because it usually leached away. After my findings I explained everything to the owner, but didn't manage to get the fertilizer programme changed or get some much needed lime.

In the meantime, the pastures were looking sick and we had hit a brick wall. I could walk across any paddock, manually pull some grass and end up with a handful of foliage, roots, dirt and all. When the cows had finished grazing each paddock, it looked like we had cultivated the area, there was so much grass being pulled out of the ground. The plants had no root depth. Then the Black Beetle arrived and got blamed for the shallow grass roots and the cultivated look. Autumn was the worst. The property was in bad shape and we couldn't do anything about it. To carry on leasing the farm was going to be too expensive. Our inputs to the system were huge as it was and production wasn't paying for them. The milk production was declining year after year and it was no wonder the landowner blamed our farming methods. We applied for another lease position advertised in the paper and were privileged to get it. We decided to make the move because our expenses were spiralling out of control.

On our departure from the Short Road property it was disappointing to learn $40,000 of our dairy company milk payments was forwarded to the landowner's account on his instruction. He believed we owed it to him. He was the shareholder in the dairy company and we were only the labourers. All of our income for the following three months had been taken from us and the only indication of the consequences was when a letter from the dairy company arrived in the mail. We were in absolute shock! How could any property owner do that?

Up until that day our relationship with the owner had been fabulous, but obviously he couldn't see we weren't at fault for the pasture's shallow root structure. It was a very painful time with shifting costs and many outstanding accounts but we had no income! Our day in court arrived four months after we'd moved from the property and the judge ruled the money was to be returned to us with interest. It was a massive relief.

Gypsy Day arrived not a day too soon and we moved our belongings, machinery and cows to a three year lease on a dairy farm, just west of Hamilton. Thank goodness our new employer understood and was appreciative of our situation. It was the first time in twenty three years our farming career had moved us away from the Te Awamutu area. It took me twelve months to become accustomed to driving and shopping in the big city. I wasn't used to all the traffic and I would rather drive the extra distance back to Te Awamutu than tackle Hamilton, besides I knew where to find everything in the smaller town. Our youngest daughter was able to complete her Agricultural Degree at the Polytechnic and work with me at the same time which was the real advantage of living close to the city.

The lease on Creek Road was supreme and reinforced our confidence as capable, productive dairy farmers. The property owner used annual soil tests as his guide to the correct fertilizer and insisted there be no urea applied to the land. We were on the same page. It was a great feeling, and he didn't mind two females running the show. The pasture was a beautiful thick sward of rye and clover and the property was completely flat. It got a little wet in winter but the stand-off pads worked well preventing damage to the pasture and providing warmth and shelter for the cows. The stand-off pads hadn't been used for a while and were in need of a revamp. At the beginning of our second season we hired a bobcat to clean them out and then added truck-loads of metal and sawdust to finish the job. It was an expense at the time, but we were rewarded with nice warm, contented cows.

The herd produced exceptionally well there. We were really proud of them. Calving 195 cows in July they produced 64,000 MS (milk solids) the first season, 71,000 MS and 72,000 MS in the second and third season respectively.

We milked in a rotary shed so drenching the girls was a challenge but we managed to continue with the same animal health programme we were used to. My youngest daughter and I made a great team and still are to this day. We know each other's thoughts and moves and frequently our anticipations have made working together run so smoothly. (I have been told by a spiritualist we were twins in a former life, which doesn't surprise me, but that's another story).

It was a huge relief to see our cows producing so well, proving the former Short Road property obviously wasn't supplying the correct environment. We were still drenching, using CIDRs and worming the young stock with a chemical pour-on every six weeks. In November early December the surplus grass was harvested into large silage bales and every year we planted a paddock of turnips. The paddocks were boom sprayed with Round Up to kill the old pasture and then ten days later the cows would go in to graze off the dead stubble. The plough would then work its magic followed by the disc and power harrow to make the perfect seed bed. Before the seeds germinated a pre-emergent spray was applied to control any unwanted weeds. It was expensive to establish but the benefit of the extra fed in summer compensated for the cost.

Thinking back to those years, I cringe at the thought of myself allowing the milking herd to graze pasture ten days after it was killed using a herbicide and then their milk being sent to the dairy factory for others to drink. That was crazy, but, it was the recommended procedure to clear the paddock of foliage before cultivation and many farmers still use the same practice. In 2014, I heard of an event where some grass in a drain was sprayed with the same product. After a couple of weeks the drain was dug out and the sludge was placed on the neighbouring certified organic property. The heaps of drain cleanings were removed just a few days later and the soil underneath had to be tested for any traces of chemical residual. The soil tests revealed the area had been contaminated. It wasn't reported as clear until six months later proving herbicides can enter the soil and take time to disappear.

The well-known herbicide Glyphosate, a major ingredient in Round Up, has recently been reported by WHO (World Health Organisation) as a possible carcinogen to humans and has been revealed in 90% of the world's food chain, even appearing in the umbilical cords of the unborn. When it was first developed Glyphosate was patented as an anti-biotic.

The owner on Creek Road was fantastic to work with. We wanted to stay on the property longer than the three seasons we signed for, but we were asked to purchase all the farm's dairy company shares to stay and we felt we weren't ready to take on that much debt. It amounted to around NZ$350,000 at the time. It was a gamble we weren't sure about and we were just climbing out of financial debt attributed to the last property. We declined the proposition and started looking for another dairy farm to lease.

CHAPTER FOUR

Discovering New Ways

Gypsy Day 2001. We were very fortunate to secure a third lease property and it was just 200 metres down the road. The property had been out of dairying for only twelve months so it wasn't difficult to reinstate and supply milk to Fonterra. The owners didn't want to continue milking cows themselves but were willing to lease the farm to us and put it back into dairying. We got on famously with the owners and they left us to manage the property as we saw fit. This we did, treating the land as if it was ours.

While living on Creek Road I began searching for *me*. I began to question what I wanted from life and just where life was taking me. I wrote down three questions I had to find answers to: what did you love as a child, what would you do if you couldn't fail at it, and who do you think you are? I found a photo of myself when I was eighteen months old, very cute, if I do say so, dressed up in a pretty pink dress and smiling for the studio photographer. To me this was a moment caught in time reflecting my *authentic self*, unshaped, free willed and without fear. From that space I concluded I was influenced, for better or worse, by my surroundings and others such as teachers, parents, and later the media and television. I conformed to how society wanted me to be and being a female my expected role was to be, a wife and mother and wherever my husband went it was my duty to follow. I had spent many years where my children and husband came first but now I felt it was my turn. I began to read a lot of books and I was *drawn* towards self-healing and personal development. I was looking for a direction in books like Weekend Confidence Coach, Power of Focus, How to Lose Friends and Infuriate People, Feelings Buried Alive Never Die and You Can Heal Your Life. I had an insatiable thirst for books and in later years they introduced me to the unexplained and spirits of our natural world. Amongst the many inspirational affirmations I discovered, one made an immediate impression, '*Do what you do best but do it better*'. I suddenly realised there was no reason for me to continue searching for a new *me* or a new direction, farming and the land was my life and it was what I knew best. I loved my job and couldn't see myself doing anything else, but I would have to do it better, a whole lot better!

In the back of my mind I had the idea of farming organically. Fonterra was offering a premium for the first three years while in conversion and I liked the idea of doing my bit for the environment. In the home we were already changing our toiletries and grocery items for certified organic brands. My daughter and I were reading the ingredient lists and looking them up in our little red 'Chemical Maze' book. There were a lot we couldn't even read and the little red book had them listed as possibly dangerous to our health. I was told if you couldn't pronounce their names, they're probably not good for you.

On the farm we were searching for something to improve the cows' health and to increase their chances of pregnancy. I stumbled upon AgriSea, an animal health tonic made from seaweed containing all the minerals a cow's diet should have. As I read the list of vitamins, minerals and amino acids in the mixture I noticed it had everything that we had been giving them, copper, selenium, calcium, magnesium plus more. We were amazed by how fast the herd's health turned around. Their coats started to shine, they were a lot happier in themselves and their pregnancy rate increased dramatically. I no longer had to give them a copper injection or selenium drench, it was all in the animal tonic in a naturally balanced form. The girls loved it, even the young stock who would fight to get their share from the drench gun as we stood amongst them in the paddock!

We were beginning to change our ways and the more we looked, the more flaws there were in the industrial agricultural methods we had been implementing for many years.

In September 2003 I attended the Woman's Health Expo in Hamilton and purchased Elaine Hollingsworth book entitled Take Control of Your Health and Escape the Sickness Industry. In her book the former Hollywood actress attacks the pharmaceutical and medical protocols documenting many statements and evidential proof involving our medical fraternity as well as hundreds of references to the dangerous ingredients in our food. I was aware of some of these facts but to visibly see them all stacked up was frightening. Elaine's book was shunned by the establishment so it was being personally promoted at the Health Expo. The mainstream distributors of health books said 'it was too far out'. The multinational controlled media wouldn't publish it and the medical establishment labelled the writer as dangerous. At the time Elaine was a director of the Hippocrates Health Centre in Australia. Her 350 page book documents her anger and disgust at the drug manufacturers, scientists and physicians who dominate the food and medical industry. It was and still is a must read for anyone looking for the facts.

'Contempt, prior to complete investigation, will enslave a man to ignorance'- Elaine Hollingsworth

I was blown away by the number of dubious food ingredients we consume on a daily basis regarded as safe by our Health Establishments of which Elaine's research suggests otherwise. Reference was made to the dangers of soy, how pasteurising milk makes it indigestible, the inescapable radiation rampage we live in, dangerous and useless vaccinations, and our deadly drinking water, just to name a few of the controversial subjects she is very passionate about. Elaine was also adamant factory farming is detrimental to our wellbeing and big is not always better. The book is presently in its twelfth publishing which notably speaks for itself.

It was around this time I had an overwhelming experience of my own. I was drenching the cows twice a day and with each dose the drench gun was leaking and dripping down the underside of my forearm. One particular day I was washing up to go to town and I lifted my arm to the mirror to make certain I had scrubbed up properly. I hadn't, so I returned to the washbasin and on second inspection the dirty mark was still there. What the hell was on my arm that couldn't be washed off? I wondered if it was grease. No, it wasn't grease but it was black hairs, forming a wide dark line from my elbow to my wrist! They were growing where the drench mixture had been dripping down my arm. After some investigation I discovered I was giving the cows a petroleum by-product. I was mortified, I couldn't believe it. I dare not mention the products name in case of retribution but I'm so glade our girls no longer get it rammed down their throat. The animal drench was being absorbed by my skin increasing the hair growth on the underside of my forearm and turning them black. I questioned my morals at that stage. Particularly around using a substance that was without a doubt making its way into the milk we were drinking and sending to the factory. I recalled when I was breast feeding, the doctor gave me a long list of what I couldn't eat or drink for the sake of the baby, so why is a lactating cow any different? I know for a fact that what they eat and drink does affect their milk.

I remember hubby visiting a dairy farm in Ohakune, the only region where the milk was produced and collected as *frothy milk*. It was used in many restaurants for their cappuccinos and the main reason the cows were producing the special *brew* was the different grasses that grew in the area. I'm also aware of bacterial injections that can be given to the cow so she produces special antibiotic milk. That's just two examples of how we can alter the final milk product and I know there are others.

Because of the dilution rates involved with the huge milk volumes processed by the dairy company a lot of substances go under the radar and do not affect the manufacturing process. It would also be too expensive to test for many of the additives administered on farm, but worse still, many feel these products are necessary to get top production. I'm horrified to admit we had also been falling into that same trap.

To survive in the dairy industry, farmers find themselves having to milk 250 to 300 cows. That was the average herd size in the year 2000. My daughter and I were milking 170 cows and as we were just a couple of women we knew we had to be the best we could possibly be, plus more!

During the first four years leasing the Creek Road property my daughter and I achieved two milestones. The first involved milking the cows once a day for the entire season and not the usual night and morning. I really enjoyed the experience because I was able to spend more time with the grandchildren and it was easier to leave one of us at home to milk while the other had a holiday. We concluded we lost $30,000 with the reduced milk production, but on the positive side, more cows got pregnant resulting in a tighter calving pattern the following season, we spent less on vet's bills and we had heavier boner cows going to the freezing works.

Their health during the season was exceptional because there was obviously less stress on them as well as us. We couldn't really put a value on that nor the extra time we were able to spend with family and friends. It was priceless. It was the one and only Christmas Day I didn't have to milk the cows that afternoon.

The second milestone was producing 420MS (milk solids) per cow. The herd's average over the years had been 320MS per cow. It was our best milking season ever and the production record gave us the confidence to manage any challenge thrown our way. We learned to *jiggle, jiggle and jiggle,* a much calmer more feminine approach, compared with the *rip, shit and bust* style of our male counterparts. Our unique technique was put into action when the occasional cow got stuck over the railings in the bail area of the milking shed. Using a rope and some gentle persuasion we would *jiggle* her out of her predicament and with an enormous feeling of satisfaction we would sit back and comment how women can do anything!

One of our biggest challenges was still to come but we were ready.

My daughter and I started to think seriously about converting to organic. We put the idea to hubby but he was dead against it. We let the idea float for a couple of months and then tried again. We were both convinced it was the right way to go. He was still dead against organics but finally offered us his conditional acceptance, 'The minute you lose cows and production drops you go straight back to the way we were'. It still wasn't a solid yes, but we took the plunge anyway.

It was going to take three years to be completely certified and we began the process in November 2006. Our introduction to BioGro, our chosen certification agency and Fonterra Organics started the journey. There was a map of the property to draw, the affidavit to sign, RMP (resource management plan) to prepare and the understanding all inputs had to be *certified organic* with the proper verification available for the annual audits.

A month before our start date we used our final herbicide to kill some gorse, blackberry and boom spray three paddocks covered with California thistles. The Californian thistle has become a huge problem invading a lot of valuable pasture and in my opinion they have spread a lot further over recent years. Spraying them that final time was the last I ever thought of them as a weed or something that had to be *killed* or destroyed. I was to learn *a plant is only a weed when it grows in a place we don't want it to be* and we should be questioning why it grows there in the first place. Weeds have a lesson to teach and a message for us to learn.

Read the message before you bury the messenger.

To my amazement I was beginning to look at the earth and my farming profession in a completely different manner discovering I knew very little at ground level. I discovered organic farming was a different world and I wanted to learn more and experience it first-hand. It was incredibly exciting and I knew this was the way we had to continue.

We said farewell to chemical fertilizers, antibiotics and a bucketful of 'cides'; insecticides, pesticides, herbicides and fungicides. We were not going to kill any more. We were going to walk alongside Mother Nature and learn her ways as much as possible.

Antibiotics was a no brainer because we had already experienced antibiotic resistance in the herd a couple of years earlier. Cow number five taught us a few things when she developed mastitis and the penicillin didn't work. We got the vet to test a sample of her milk and our worst fears were confirmed, no antibiotic could help her. We milked her once a day into the test bucket and discarded her milk. Our plan was to send her to the freezing works as soon as the meat withholding period for the penicillin had expired. To our surprise three months later she had recovered from her mastitis and her milk was included in the herd's supply to the factory. The following month we tested her milk along with the rest of the milking herd in our tri-annual herd test. Number five's somatic cell count, (a general sign of mastitis) was so low it was barely countable. She had cured herself completely and all we had done was given her body a chance to heal. Mastitis is a sign of low immunity and like many illnesses and diseases they will strike when the body cannot defend itself.

With our change to organics we had to find an alternative to antibiotics and we were introduced to homeopathy, probably not a moment too soon. We went cold turkey and have never regretted the decision. Homeopathy is a simple concept with *like cures like* but without the freely given guidance from Homeopathic Farm Support in Hamilton we would never have made the grade. All we had to do was ring them, describe the sick animal and they would post the solution which would arrive with the mail the following day.

Homeopathy is based on an energetic signal, an information signal. You choose a substance that copies the illness, not one that opposes it. The substance becomes increasingly potent as a portion of the illness is added to alcohol and energised by the practitioner. The solution is diluted and energised at least twelve times after which the Avogadro number is passed and there is no original substance left. The Avogadro number refers to the number of units in each mole such as the electrons, atoms, ions or molecules. When you set up a spectrographic transmitter on the solution, you find the more you dilute it, the more it starts transmitting energies. The spectrograph is the most sensitive scientific instrument available. It can even tell you if there's manganese on a star that's a thousand billion light years away!

To be able to administer the correct homeopathic solution the correct diagnosis is very important. Ultimately we've found ourselves reading the animals with so much more empathy and awareness of the situation that requires treatment. For mastitis we now had several solutions we could use depending on the description of the mastitis at the time. There was no 'one jab fits all' regime like we were used to. We now had to look and feel all the symptoms: her temperature, what side she laid on, was she eating, was the udder cold, how was she walking, did the symptoms appear suddenly? We learned to paint a picture of the disease and look at the cause as well as symptoms. And then there was the new vocabulary to learn. The names of the solutions were very foreign to us in the beginning. Tuberculinum, Nux Vormica, Pyrogenium, Bryonia, Hepar sulpuris, Colchicum, Folliculinum, and there were many more like these that would come to play a huge role in the herd's health as well as our own from that day on.

Chemical fertilizers had not played a huge role in our most recent farm management but to find a suitable fertilizer that was certified organic wasn't easy. The two major fertilizer companies had nothing to offer so we had to search elsewhere. We had been using AgriSea Animal Tonic with impressive results and decided to use their seaweed fertilizer. The bonus here was their product was certified organic, easy to apply and readily available to the plant because it was in a liquid form. I noticed the colour of the cow pats changed to a deeper green after our animals grazed the pasture where it had been applied, so I knew the nutrients in the AgriSea were moving through the food chain. We were satisfied with our organic farm management but we were curious to know if we were on the right track, after all hubby and the neighbours were watching our every move.

I made enquiries and management at AgriSea recommended Dr Greg Tate who they felt knew what he was talking about and basically other than him, there was no one out there. Dr Tate wasn't affiliated with any fertilizer company and had recently walked away from a highly paid position where he had been inventing agricultural chemicals. He told us, 'One day I saw the light and remarked to my fellow scientists, I think we are going about this the wrong way.' His colleagues thought he had gone bonkers, but he had come to realise *the more of Mother Nature we try to destroy, the stronger her defences will grow. We kill one species and another will take its place.*

Dr Tate (or Dr Greg as he is known) paid us a visit and did a farm walk with us. The first paddock we went into he said 'This place needs calcium'. I was very surprised, we had just put ten ton of lime on that particular area six months earlier. Dr Greg said the soil should be spongy to walk on, not crusty and hard. He took some soil tests and said he would do a Reams Test as well. Dr Greg explained the Reams Test is used to show what minerals are available to the plant whereas the traditional Hill's Soil Test will indicate what is in storage. At the time the Reams Test had only recently been invented by Kerry Reams, so we had to send our soil to America for testing. Meanwhile Dr Greg made enquiries at all the fertiliser companies looking for an organic mix that would suit our situation. In his opinion the best fertilizer available was made and marketed by Grant Paton of Environmental Fertilizers in Paeroa.

We respected his recommendation and the fertiliser recipe was put together for us. The farm was to have two fourteen ton applications of solid fertilizer every six months plus a liquid mix to go on a little and as often as possible. The solid mixture arrived, still warm and active with plenty of steam coming off it. It was the first time in my life I had ever seen a dog eating fertilizer and she *tucked in* as soon as it was tipped off the truck. She absolutely loved it! The fish based fertilizer had a distinctive smell and obviously a yummy taste for canines. Come to think of it, from that moment on, we took more notice of what our dogs ate and this included watching them enjoy freshly dropped cow pats. They didn't always go for them suggesting a tasty cowpat to the dog only comes from a healthy animal having eaten healthy nutritious pasture.

The solid fertilizer was applied by truck using the broadband spreader and we spread the liquid mix from the four wheel motorbike with a rosette sprayer on the back. It was about that time I began to look at our pasture in a different way and I questioned my feelings and empathy towards the growing plant. If it was me being eaten to ground level I'd be wanting food to give me the energy to grow, and what about after a cold, wet day or a hot, dry summer wouldn't I be feeling hungry and lethargic? I know I wouldn't survive on just two meals a year. Who made that rule making us believe plants want food only twice a year? Dr Greg suggested a little and often is the best recipe.

If it looks hungry, it will be, so simply feed it.

Dr Greg put the liquid mix together using Environmental Fertilizer's ingredients. I honestly cannot recall it completely but I do remember the liquid calcium and the seawater! Yes, seawater gathered from the purest ocean beach we could find or off a fishing boat out in deep water. Many have questioned us saying seawater will kill the grass. I assure you, it does not kill anything. The sea is where all life began and holds every mineral on the periodic table and probably more! Compare the taste of your own blood with a teaspoon of sea water, you may be surprised. They taste the same to me.

The cells in the body of an older fish in the ocean have exactly the same structure of those in a baby one. Cells do not age in seawater. That suggests to me, ocean creatures are being nurtured in the ultimate environment. An environment full of the nutrition required for life to exist so why shouldn't we look at applying it to our soils. Next time you visit the ocean, take home a twenty litre container of seawater and tip it out on the lawn. The grass will not die, it may show signs of accelerated growth and become a darker green, suggesting an elevated level of photosynthesis, but you will not kill it. Charles Walters examines Dr Maynard Murray's career and success behind applying seawater to crops in his book Fertility from the Ocean Deep and it makes inspirational reading.

When we first applied seawater to the pasture, we instantly recognised the milking herd was more relaxed and contented in the shed. It didn't matter when it was applied, we still noticed a change in the animals' attitude and the different colour of the pasture. I don't recommend using it during the heat of the summer but any other time of the year is okay.

Calcium will always be an important part in any fertilizer programme and by applying it in a liquid form the mineral is immediately available to the plant. Calcium is the only mineral the plant cannot harvest from the atmosphere. Each fertilizer programme will be tweaked according to recommendations and soil tests, but I have come to the conclusion no one knows exactly what a plant or soil requires. In my opinion it's better to give a balanced food with all the ingredients and allow the pasture to convert what it can use and discard the rest. The fertilizer mix from Environmental Fertilisers was exactly that, a complete well-balanced food for the soil and microbes.

We documented how the mineral ratios changed during the time we were organic on Creek Road with four Reams Tests. The property was split into two halves, the centre and the swamp and the results can be found in the appendix.

The Reams Tests showed a gradual increase in most areas which is amazing for the short twenty month period recorded. The Potassium count is indicative of the mineral being released after being 'locked' in the soil and is now being flushed out. Note here that we produced 420kgs MS per cow in the 2005/2006 season, the one prior to the 2007 tests which confirms far too much emphasis is placed on scientific reports. They are a guide only and not something to lose sleep over. I once witnessed forty different soil compositions in a fence line where we had manually excavated forty post holes suggesting no two areas of earth are mirrors of the first.

So unless each soil report is taken from exactly the same spot with exactly the same weather conditions, exactly the same time of day, variations will suffice. You also have to have the exact same cosmic forces working because certain minerals are drawn closer to the earth's surface as planets move within the universe. We can't expect not be influenced by the multitude of natural forces in the skies above us.

On August 10ᵗʰ 2007 we also took a Hills soil test from the same areas to investigate the level of minerals stored in the ground. These results can also be viewed in the appendix.

Over the years scientists have dabbled with soil tests at the farmer's expense. I have collected soil tests from every property we have worked on and for the following comparisons I have chosen the years 2007 and 2012 which show how the medium levels recommended by scientists have declined.

The following are the medium ranges from our Hills Laboratories soil tests, the 2007 year on the left and 2012 to the right.

2007 Magnesium 1.00 – 3.00 2012 Mg 1.00 -1.60
2007 Potassium 0.50 - 0.80 2012 K 0.40 -0.60
2007 Calcium 6.0 – 12.0 2012 Ca 4.0 – 10.0
2007 pH unit 5.8 – 6.3 2012 pH unit 5.8 – 6.2

Every farmer likes to see their soil tests showing the above minerals in the medium range and in some instances higher, so why the change and what does it really mean? Can't we get our soil levels as high as was first suggested or have we applied too much of these macro minerals? Have the ranges been altered before and will they be changed again without us being informed? This is one reason why I felt we had to apply a balanced fertilizer containing as many minerals as possible and allow the plants to take what they want. I have lost faith in *the system*. I also knew the fertilizers we were dealing with were *alive* making them instantly available as food for the pasture as well as the soil microbes.

Science is known for constantly *moving the goal posts* when it suits. Unfortunately it will take 400 years for scientists to investigate every mineral component on the Periodic Table and discover the ultimate recipe for our soils requirements, we simply do not have that long! Our soils are nutritionally depleted now and so is the food we harvest from them. It's a known fact some minerals have depleted by as much as 64% since WWII. In my opinion science is leading us like lambs to the slaughter and time is running out.

It doesn't always pay to follow the crowd.

I read with amazement David Montgomery's book entitled Dirt, The Erosion of Civilizations. He documents how the earth has been raped by farmers and early settlers since the beginning of time. How mankind has demolished forests and virgin landscapes to their detriment as the earth has been left scarred and open to the elements causing erosion and degradation of our precious soil. So much of our soils are ending up in the ocean, and this has been occurring worldwide for many years. In some parts of Gisborne in New Zealand, 25,000 tonnes of sediment is washed off each square kilometre every year. Very few areas are immune to soil degradation and loss of topsoil. There is a lot of highly erodible hill country in areas such as Whanganui in the Manawatu region, Hawkes Bay, and inland Taranaki.

My organic journey received a huge boost when I read page 207 of Montgomery's book, in the chapter Dirty Business, and I quote: *Long term studies show that organic farming increases both energy efficiency and economic returns. Increasingly, the question appears not to be whether we can afford to go organic. Over the long run, we simply can't afford not to, despite what agribusiness interests will argue.......A number of recent studies report that organic farming methods not only retain soil fertility in the long term, but can prove cost effective in the short term.*

David Montgomery also makes reference to the research led by Washington State University's John Reganold back in the 1980's where he compared the state of the soil, erosion rates, and wheat yields from two farms near Spokane in eastern Washington: *One farm had been managed without the use of commercial fertilizers since first ploughed in 1909. The adjoining farm was first ploughed in 1908 and conventional fertilizers regularly applied after 1948. Reganold was surprised to find there was little difference in the net harvest between the farms. From 1982 to 1986 wheat yields from the organic farm were about the average for two neighbouring conventional farms. Net output from the organic farm was less than that of the conventional farm only because the organic farmer left his field to fallow every third year to grow a crop of green manure, usually alfalfa. Lower expenses for fertilizers and pesticides compensated him for the lower yield. More important, the productivity of his farm did not decline over time.*

Reganold's team found and I quote: *that the <u>topsoil on the organic farm was about six inches thicker</u> than the soil of the conventional farm. The organic farm's soil had greater moisture-holding capacity and more biologically available nitrogen and potassium. Soil on the organic farm also contained many more microbes than the conventional farm's soil. Topsoil on the organic farm had more than half again as much organic matter, as topsoil on the conventional farm. The organic fields not only eroded slower than the soil replacement rates estimated by the Soil Conservation Service, <u>the organic farm was building soil.</u> In contrast the conventionally farmed field shed more than six inches of topsoil between 1948 and 1958. Direct measurements of sediment yield confirmed a fourfold difference in soil loss between the two farms.*

As I read Dirt, The Erosion of Civilisations I found myself agreeing with the author with astonishment and fortitude. Astonishment that the human race had the ability to learn from mistakes of past generations, but nothing has changed. There has been no lesson learned, and fortitude, in that I write this book with my heartfelt plea for readers to stand up and be the change, before all is destroyed.

CHAPTER FIVE

Beneath My Feet

My respect for the amazing world beneath my feet was enhanced tremendously when I participated in the ten month Organic Horticultural course with tutor Micky Cunningham in Hamilton. The year was 2009 and the course was recommended to me by Brenna, our veterinarian at the time. The class was limited to twelve pupils, was a once a week event and a NZQA accredited course. There were subjects introduced that every farmer in the country should be made aware of and I felt so dumb in not knowing. One aspect that really shocked me was the single request to draw our favourite plant. I recall thinking 'Okay, that's easy, what's the catch?' My drawing was of a pansy plant with flower and foliage. Eleven members of the class drew similar pictures of personal favourites. Only the twelfth participant got the image correct. It had to have all the presentable criteria as is visible on the earth's surface, but then complete the picture with the plant roots reaching down into the soil.

I had never thought seriously about what goes on below the earth's surface. Yes, I had studied biology but until I was faced with the practical aspect I was only writing *facts* on paper and passing exams! My lessons were about to commence and it was the most awe inspiring time I had ever spent in a classroom. The following is an article from a Resource booklet made available to the class where the question is asked 'What is Soil?' The following breakdown of the soil's ingredients and its functions were monumental to the lessons I had to learn.

25% Water
Dissolves & carries nutrients and other minerals to plant & soil

25% Oxygen
Essential for soil health
Soil should aerate like a 'lung' due to air pressure
Essential for root growth, water & nutrient uptake, & biological activity
Anaerobic means a lack of oxygen (water logged)
Roots & micro-organisms respire needing oxygen for photosynthesis

45% Earth
From weathered rock, broken down by the elements
Used by all living beings, recycled and reformed
Water & air is 50% of the space in soil and called soil pores

5% Organic Matter
A valuable food source for all the soil bugs

<u>*And 'Fire'*</u>
Without the farmer's enthusiasm, interest, energy and love of their work, all other elements won't ignite!

It was awesome to see the importance placed on the farmer's role. For this dumb farmer I hadn't equated for the oxygen content in the earth I walked on, nor had I considered the number of micro-organisms I trampled with each step. In just one teaspoon of earth there are more inhabitants than there are human beings on the whole planet!

While many consider the earthworm to be king of the soil, the smaller microbes are just as important. Unfortunately our underground inhabitants are very vulnerable to the damage done by modern farming techniques, heavy machinery and especially chemicals, urea and soluble fertilizers.

I have discovered a simple method of testing the soil for life. One of my favourite outside jobs is chipping thistles, especially the Scotch thistles who grow with large tap roots. A clean cut to the main stalk a few millimetres below the ground reveals the ultimate area to test the earth. I pick up a handful of the disturbed ground around the cut stalk and smell for the *aroma*. It is so, so beautiful. Such a sweet, earthy perfume. That is what soil should smell like. If I picked up some soil elsewhere in the paddock there's a 99% chance I wouldn't find the same aroma. The amazing Scotch thistle and his friends use their tap roots to reach down into the earth and draw up minerals that feed the microbes nearer the surface.

To me, that *aroma* is indicative of good health and I have discovered there is a comparative scent that accompanies the healthy bovine, a healthy dog and believe it or not, a healthy child. Amazing!

Another of my discoveries involves being able to whip the cream off the top of our four litre milk billy. Years ago I tried to whip some cream from our cows without any luck. I decided to leave it in the fridge for a couple of days and try again but it still wouldn't whip. The result was always the same, so I gave up. A quick trip to the supermarket was so much easier. Many years later as we moved away from our conventional methods and into organics I got a wonderful surprise. I gently scooped a cupful of cream from the top of the milk billy, collected twenty four hours earlier, put it into a bowl, and started the electric beater. Within four minutes the cream was whipped! Absolutely incredible! Our cows have always been predominantly Friesian and Friesian cross breeds whose milk has a low fat content, so how was it possible? I compared my whipped cream success with the neighbour's jersey cows who always produced creamier milk and a higher fat test. The wife said it was impossible for her to get whipped cream without storing it in the fridge for at least a couple of days. And then she said she couldn't get any cream to whip before December when their milkfat tested above the 5.00. I was astonished because here I was in August whipping this beautiful cream with a 4.03 fat test and I could do so within twenty four hours of it being collected. Organics had definitely changed our girls for the better. Since my discovery I've made heaps of homemade butter by leaving the electric beater on until the whipped cream becomes a thick blob but mostly the cream is too delicious to let it go to the next stage.

My initial step towards farming organically was an uphill climb. I attended as many organic Fonterra discussion groups as I could and from each one I'd gather small pieces of information to help my purpose. I noticed no two farmers and no two properties were alike. We all had our own individual recipes that were tweaked along the way.

One of these discussion groups took us to the Ruakura Research Station. The dairy manager took us on a tour of the property on the back of a couple of four wheel drive vehicles. As far as the feeding and grazing management was concerned it was just like any other farm but it was unique in other aspects. It was situated on prime Waikato soil on the outskirts of Hamilton. At the time of our visit there were twenty seven cows on the milking platform with a smaller number participating in a nutritional experiment where an open vent to their gut provided immediate access to their rumen. As a group we found it difficult to comprehend how the small number of animals in such a prime location and environment could provide unilateral data for New Zealand's farmers. I also felt their research was in *ground hog day*. I had recently read an old farming magazine where similar projects had been investigated two decades ago!

My organic classes in Hamilton were always a day to look forward to. Early in the course we put together a compost heap on a student's property. It was quite a procedure. The carbon and nitrogen compounds had to be in the correct ratio otherwise the end result would have been disastrous. Two thirds carbon to one third nitrogen and each layer added to the heap was significant. We began by disturbing the ground where the heap was to be matured and then placed a layer of sticks down to allow air into the compost. Then we began to add the ingredients, one layer at a time beginning with the carbon: straw, sawdust, hair from cow's tails, dried leaves, and then a little sprinkling of water for moisture before adding a thin layer of lawn clippings, a little cow dung and some chicken manure to help activate the heap. Together we counted over a hundred layers in the heap. Now and again EM (Efficient Micro-organisms) were sprayed over it and a minute sprinkle of fine lime. As the heap grew two poles were inserted to be built around so at completion they could be removed to revel a couple of holes for the air to circulate. The compost was covered with old carpet and left to mature but had to be turned every few days and we recorded the temperature it attained. If it didn't reach the required heat the compost wouldn't *cook* properly. We put a dead chicken in the mix as well and never saw it again! The microbes broke it down and left it as compost. There are many ways to make compost. Biodynamics has six preparations to add to the recipe for their heap but the secret is to get the carbon to nitrogen ratio correct to ensure success.

CHAPTER SIX

My Dreams are shattered

My organic journey was about to get more complicated and make a complete U-turn as we were nearing the end of the three year term to qualify as certified organic. To me having certification indemnifies our dedication to providing the best possible product and also shows our determination to nurture our environment as nature intended. Six months before we reached full certification, the Creek Road farm we were working on had a change of ownership and the new owner was going to take it out of dairying and plant the entire area in maize. As much as we tried, we couldn't raise enough capital to make the purchase price. It was a huge knock. With everything going so well, we had taken the lease for granted and now we were forced to find another property to milk our girls on. This also meant we had to re-start our certification. If we'd been fully certified we would have lost it, and not been able to re-certify, those were the rules. The biggest misfortune at the time involved repaying Fonterra all the premiums we'd earned in the past two and a half years, a total of $NZ35,000.

During our time on Creek Road we leased a forty acre block on Perkins Road, a couple of kilometres away to graze our young stock and we included it on our organic certificate. We didn't want to lose that as well as the milking platform so our search for the next property had to be a local one. We made contact with a Willow Road family who wanted to lease out their dairy block of sixty eight hectares. We explained how we wanted to continue under organic principles and they were more than happy as long as it didn't mean applying urea. That definitely wasn't a problem.

Gypsy Day 2009 was approaching and we had everything packed ready to move. We could shift the furniture and farm machinery ourselves and the 160 cows were booked in with the trucking firm. Unfortunately everything came to an abrupt halt because the house we planned to move into wasn't how we thought. On the 28th May, four days before the big move, the previous tenants finally moved out and we had our first look inside the house. We were all close to tears as we walked from one room to the other, it was disgusting. It bought back memories of Taranaki and the challenge my parents faced. Hubby said 'There is no way we are moving into that!' We knew the section around the house was rough but we didn't expect the inside to be like it was. It was unbelievable. There was old torn carpet in the lounge, dining and passage and it was alive with carpet beetles. We could see them jumping everywhere! Then we saw the bedrooms. The first one was partially painted in an erotic purple colour, window frames and all. The second bedroom was an iridescent green with spontaneous messages written on the walls and doors. Bedroom number three was cold and dark, made more so with the brown paint on the walls. Any curtains left hanging in the windows were torn and covered in mildew. Then we saw the shower cubical, well past its use by date with a hole in the wall, covered in mould and both taps ripped from the wall. How much worse could it get? The kitchen, with broken dark green cupboard doors, mouldy cupboards and fly spots covering the old creamy coloured walls and ceiling. No, we were not going to move into that, but what the hell do we do with only four days until the shift?

We spoke to the owners who suggested they would lay some new carpet in the lounge, dining and passage way. I put it to them that if they provided the paint, we would paint the house from top to bottom before the carpet went in. It didn't seem plausible to lay flooring when the place had to be cleaned and some fresh paint applied to the walls and ceilings. Thankfully the purchaser of the Creek Road farm gave us an extra week free of charge to get the Willow Road house in order before we had to move out of his house.

It was all hands on deck, members of our family, as well as representatives of the property owner. All the walls and ceilings were sanded and given an undercoat. As we painted an area, the heaters were used to speed up the drying process ready for the second and third coats. The carpet was ripped up and chucked out on the lawn, leaving the plastic underlay on the floor boards ready for the new one.

The kitchen was a real demolition sight where every cupboard was scrubbed, painted with an undercoat followed by two layers of fresh paint. At one stage there were five people working in a very small area. In total there were twelve of us giving the house a face lift. I also had to call on my brother, a self-employed plumber because the toilet didn't work and there was trouble with the water inlets to the house plus we had to make a space in the kitchen for the dishwasher. The mould he found behind the kitchen cupboards and later in the old shower area was extremely dangerous so we gave the walls a good spray of chloride lime hopefully killing the worst of it. We also had to spray insecticide on the floor boards where the carpet beetle had been so they wouldn't re-infest the new flooring.

By June 5th we had painted the lounge, dining room and three of the four bedrooms. It was beginning to feel and smell like a home. We started to move the furniture in on June 6th. The cows and machinery had arrived on June 1st and it wasn't long before we realised the farm was also in need of some TLC (tender loving care). The owner had already got the cowshed up and running after being out of circulation for the previous eighteen months but beyond the cowshed there were many fences laying on the ground with no electric current to keep stock in. As we shifted the cows from paddock to paddock we had to repair the surrounding fences to keep them under control. There was a deep gully around three quarters of the boundary which offered a great barrier zone between ourselves and the neighbours. Great for our organic certification, keeping our side free from any unwanted chemical sprays but to our amazement the gully wasn't fenced off and if an animal went down there, there was no return.

After losing two cows to the gorse and swamp at the bottom of that gully, we took it upon ourselves to erect a two wire electric fence right around the area. We hired a fencing contractor and purchased the posts and wire for the job. You may be thinking that's a bit over the top, but we had already lost $4,000 in two high producing cows, so what was another couple of thousand to prevent it happening again. In total two and a half kilometres of fence was erected in five days and that fence saved us a lot of time and money in the years to come. Time, in that we didn't have to spend it searching the gully for animals, and money, because we didn't lose any more!

We worked really hard on the Willow Road property. The gorse had been taking over fence lines and it was ten foot high in places. All of it had to be cut down by hand to get the electric fences back in working order. We were farming under organic conversion so using chemicals was out of the question. The freshly cut gorse stumps were plastered with salt to stunt any new growth and it was gradually pushed back into the gully. With the gully fence erected and the gorse cut back we planted flaxes, willows and Tagasaste to grow as a barrier and keep the gorse under control. Tagasaste is a nitrogen fixer like gorse and grows well in similar areas. It is also a great stock food, really great in dry summer conditions.

The section around the house also got a major

makeover. The old hydrangea bushes were eight to nine feet high and falling over onto the lawn area by at least five metres. An old climbing rose bush had taken over a silver birch tree at the entrance on the roadside and was a real mission to get rid of. The kiwi fruit and grape vines had not been pruned for many years and were covering large areas of garden and growing out of control up into the native trees. It was quite a mission but after a few months we were proud to call it home.

The property was our biggest challenge but the next was somewhere to send our organic milk. The bank wouldn't support us to buy the shares required to supply our milk to Fonterra and at that time farmers were asked to pre-pay the complete amount to cover the coming season's production. Less than twelve months after our dilemma Fonterra changed its policy and allowed the purchase of company shares to be spread over a three year period.

We finally arranged for our milk to go to the Organic Dairy Factory in Okato, Taranaki. It gave us extreme pride to be sending our produce to an organic factory where they made the best cheese we had ever tasted. We were supplier number 204 and organic milk from as far as Northland in the north and Manawatu in the south was being collected. That was the factory's first season in production and everything started with a rip and a roar. By the beginning of November a few cracks were beginning to show and for some reason there was a split and conflict of interest amongst management on how the company was to function. By January our milk cheques were stopped. We were devastated. Our milk was still being transported to the factory, but we weren't being paid for it. The conflict amongst management didn't improve and the bank put the company into liquidation affecting many farmers, supporters and shareholders alike. It was a very stressful and emotional time for all. The initial plans for the factory had taken years to put together and then to get it built and processing had been a monumental task, but it wasn't to be.

Back on the farm we managed to have our milk collected by Green Valley, a small privately owned enterprise, and thankfully got some money in but they couldn't do it long term. Two weeks later Fonterra came on the scene and announced if 75% of the organic suppliers changed over to them, all of our milk would be collected. Some farmers had had enough and were not Fonterra supporters for the way they had treated organic suppliers in the past so they moved to supplying other smaller companies. Some gave up dairying altogether, their dreams and passion shattered.

For us, we didn't have much of a choice. We still had lease payments to make and the bills were stacking up. There was still the problem of 'You must own shares in the company before you are permitted to supply'. It was at this time Fonterra came up with a scheme to purchase the respective shares and pay them off over the three years. They never did get 75% of the organic suppliers they demanded and they made us wait before offering their proposed solution. In March we became Fonterra suppliers once again.

The Organic Factory had given organic farmers a chance to have their own identity and control over the premium paid for their milk. For years these same entrepreneurs had seen New Zealand as the ultimate place to expand the production of their unique product and to become the only country in the world to do so, but their ideas had fallen on deaf ears. Fifteen to twenty years ago the government had a chance to develop New Zealand into an organic paradise, but they voted against the proposal. Now we are seeing the small farmer ostracised by big landowners and pushed out of existence as a more unethical style of agriculture has evolved. The bigger properties have a large number of Fonterra shares which effectively gives them several votes on company issues whereas the small farmer is generally only entitled to one. It is no longer one farm, one vote.

We had just been through a cruel and unforgiving twelve months which resulted in the loss of $NZ130,000. Others lost a lot more. We virtually had six months without an income. It was a summer drought as well but in a way that turned out to be lucky for us. Once the government declared the Waikato a drought affected region, we were able to go to WINZ (Welfare Institute of New Zealand) fill out some forms and be given some grocery money from the drought relief fund. This was very demoralising to say the least, but we had to do it. I consoled myself with the conclusion that I had fed thousands of families with the milk we produced over the years and it was now time for them to feed me.

In the midst of all the upheaval, five months after our

move to Willow Road our daughter's son Ethan was born. Having baby Ethan in the house kept us all smiling and made life worth living. He gave us the biggest reason ever to fight on and continue our organic passion. He was a fantastic distraction to have around, cute and cuddly too.

We learned some life inspiring lessons during those lean times. Our number one priority was to take care of our animals and the land and thanks to the organic methods I had learned we managed both at very little cost. Homeopathy was used for all our animal health requirements while the land was fed using liquid fertilizers such as EM (Efficient Micro-organisms), fish fertilizer, comfrey, biodynamics and seawater.

I was introduced to EM while attending the Organic Horticulture classes and it has become a mainstay of my journey into the organic realm. I have used it as a fertilizer, in the sludge pond to assist aeration, on a tree covered in lichen and on our Labrador's facial lesion. The latter was the most remarkable. Tess had this horrible looking growth on the side of her cheek. My daughter and I took her to the vet and a sample was taken for testing but the results provided no explanation as to what the growth was. Tess was prescribed antibiotics which my daughter refused to give her, saying 'She isn't sick!' Instead we sprayed the lesion with EM for a couple of days and to our relief and amazement the growth started to recede. Within seven days we could see small hair follicles growing and her cheek began to heal extremely well. The vets got the biggest surprise when we returned the antibiotics for a refund explaining they weren't required and the lesion had healed over.

The lichen on the tree was also a remarkable event. I sprayed EM covering as many limbs and branches as I could reach. It was obvious the tree was struggling as only one third of it was growing leaves. The following spring there was a dramatic increase in the amount of young foliage and the tree looked like it had a new life.

As a fertilizer, the most visible region to see EM at work is in a paddock where water has been unable to drain away creating a stagnant, green ponding area. EM will enliven the soil and increase its drainage capabilities. I have no doubt EM provides a positive response no matter where it is applied. It is a living culture capable of activating the *good bugs* to neutralise any *bad bugs*. It is very simple to make and requires no special equipment to do so. Another area where we have incorporated EM is as a teat spray on the cow's udder after milking and we have been using it in this manner since 2007. EM was introduced to NZ by Naturefarm who are based in Christchurch. They are responsible for publishing some amazing data proving fodder crops will produce a third more where Efficient Micro-organisms is applied at the recommended twenty litres per hectare.

Seawater, fish fertilizer and comfrey tea didn't cost us a cent and were irreplaceable, especially the fish mixture which was used as a nitrogen and mineral booster for the spring growth. The golden rule here was to always add EM to the fish fertilizer ensuring any bad bacteria that could potentially harm the cows was neutralised. The other rule I had was to let the drums of fertilizer mature for at least six months before we used them. After that time frame the brew matured and developed a beautiful pink oily colour. The fish fertilizer attracts honey bees to the motorbike as it's sprayed out and the cows love to lick the residue left on the bike's mud flaps. A small rosette on the back of the four wheeler sprays the mixture out as a fine mist but it occasionally blows back onto the operator. The smell is unique to say the least and it repels workmates and spouses for at least twenty four hours!

All of the paddocks were fertilized at least six to eight times each year. The Willow Road property had an issue with the invasive Californian thistle and large areas of fleabane both indicating soil compaction. It wasn't going to be a quick fix. The pedometer readings confirmed the compaction, probably caused by excessive cropping in the past and there was a solid pan where the plough had turned the ground over. I later learned the property had been in maize for several years before it was returned to dairying. The damage was very visible after heavy rainfall when the puddles didn't drain away as they should have but by the third year we were seeing real improvement. Every season we had one particular paddock that grew a *crop* of yellow bristle grass which was becoming a major challenge for ourselves and the surrounding district. Another challenge was when the clover fleas invaded the pasture. The clovers didn't stand a chance as they were gnawed to the ground. We really had to work hard for our money on Willow Road.

My introduction to Biodynamics came midway

through my Organic Horticulture course in Hamilton but I felt it was too far-fetched for me to understand and I wasn't ready for it. I'd also visited a couple of dairy farms where it was being practised where I didn't like the results I saw. That was all due to change.

In the autumn of our second season on the Willow Road property Glen Atkinson paid us a visit and gave us his projection of the philosophy behind biodynamics. His interpretation involved the whole cosmos and how each star and planet draws and retracts electromagnetic fields. We all know how the moon affects the earth's tides so it is only rational to suggest we are being influenced by the solar systems electromagnetic fields 24/7. I was in awe of how he explained his research but I was still not ready to fully comprehend the lessons. On Glen's departure he gave us a free sample of his fertilizer product called PhotoMax, one of many under the trade name BdMax. Hubby applied the product to three paddocks. The *proof of the pudding* came four weeks later in late May.

We were milking six cows over the winter months that year and for a short period of time our daughter noticed the house milk had a strange taste. Something wasn't right. I started to blame our cleaning procedure in the dairy shed and I became really concerned because we were adding the raw cow's milk to Ethan's milk mixture at the time.

The six cows had been doing the rounds getting a couple of days in each paddock for the duration of the tainted milk episode. We shifted them to where the PhotoMax had been applied and as soon as the cows walked through the gate they began to graze. They didn't rush to the back of the paddock as we had been seeing, they decided the entrance way was sweet enough. Believe it or not their milk was so much creamier with a delicious flavour and the strange taint had completely disappeared. The transformation was so noticeable it could only be attributed to the PhotoMax we put on the pasture, nothing else was changed. We later put them back into the other paddocks as an experiment and sure enough the tainted milk returned. I never did find a reason for the tainted milk but I suggest the PhotoMax changed the pastures composition. From that moment on I was sold on biodynamics and my understanding of the concept has increased enormously.

Another of the biodynamic products we used was called Etherics 1000. Etherics 1000 is a general health tonic incorporating all the biodynamic preparations and I was told it could provide a good control against many insect problems. I decided to see if it could help us with our clover flea problem. The fleas were so thick there were hundreds jumping on to my gumboots as I walked across the pasture. We sprayed the biodynamic preparation on two paddocks as an experiment and within two weeks we were seeing young healthy clover leaves growing again and I couldn't find any fleas. I had doubts about the result so I compared the treated paddock with the one next door. Maybe it was time for the clover fleas to move on and they had just disappeared. To my amazement they were still in the adjacent pastures devouring the clover and there were just as many jumping onto my boots as there were before we applied the Etherics to the two paddocks.

Biodynamics is seen as the ultimate organics, a world influenced by the many forces far beyond the earth itself. It takes into consideration the interconnection of the universe as a whole with each element directly affecting others. Rudolf Steiner, back in the 1920s was the philosopher who introduced the biodynamic principles to Germany. He was a very inspiring scientist, well before his time who also foresaw today's declining bee population and the fertilizers applied to farmland not being the correct composition to sustain our soil's health for the generations to come. His philosophies took into account what Mother Nature naturally provided and he administered the biodynamic preparations accordingly.

The BdMax products played an important role in our fertilizer programme from that time on. I began to add the biodynamic preparations in their raw state to my 200 litre drums of homemade fertilizer as well. The preps were numbered by Steiner beginning with 500, which is cow pat in the bovine's horn. 501 is silica, 502 a yarrow preparation, 503 chamomile, 504 stinging nettle, 505 oak bark (a rich source of calcium), 506 dandelion and 507 valerian. Each of the listed herbs plays an important role offering the appropriate mineral to the soil and plant when the *brew* is ready. For me, I can't decide whether it is pure magic or the perfect recipe. Either way I love the results.

While living on the Willow Road property I attended a two day course just outside Morrinsville featuring Dr

Arden Anderson and Dr Paul Dettloff from USA. Dr Anderson was a practising General Practitioner and qualified soil scientist and Dr Paul the veterinarian in charge of several organic dairy farms in his home state. They provided us with an amazing insight suggesting how we have all been duped to follow the American way. They both passionately begged us to not follow *the piped piper*. Dr Anderson was seeing the results of poor nutrition in his patients and quoted 98% of all illness is due to that fact, and good nutrition starts with healthy soil and healthy soil relies on the farmer.

Day two of the seminar took us on a bus trip to the Hauraki Plains where we visited a perfectly set up dairy block but it had an enormous problem with the pasture being ravaged by large black crickets. The area is renowned for their enormous population and we could visibly see them jumping out of the cracks in the ground as we walked along. We saw them in their hundreds that day. The suggestion was put to the owner that he should change the *environment* to one where the cricket didn't want to live and breed. It was obvious the chemical sprays were not working 100% and it was only a matter of time until the cricket would become immune to them anyway. Both of our American visitors had seen it all before in their own country. The group discussed the farmer's dilemma and as a group we felt we made an impression but only time would tell if the suggested change to the environment would be instigated.

Another of their suggestions was to increase the variety of pasture species on the property. A bovine should have at least one hundred different plant species available to them every five days for optimal health. Wow! The hundred should include trees such as the willow which is high in calcium, the oak, native flax, feijoa trees full of iodine, and herbs like self-heal, chick weed, dock, buttercup and yarrow to name a few. I was amazed at the concept and not long after the seminar I put the scenario to the test with a group of school children. On their farm visit I gave each group a small bucket and asked them to collect as many different leaves as possible while they walked from paddock to paddock. An hour later we had a count up. There were only forty eight different plant species, well short of the hundred I was hoping for. We had a long way to go.

Years ago there would often be *the long acre* where a sick animal could graze amongst the scrub at leisure and be able to self-medicate. The animal would have a varied diet of medicinal herbs and trees, each with a different mineral content. It is a very different story on today's modern property where the farmer is encouraged to sow just rye grass with a couple of clover varieties and the farms have very few trees. Chicory, high in zinc and other minerals has recently become the fashionable herb to sow in the pastures and plantain is another. Let's hope there are others to follow.

My home library was making a huge transformation about that time. I began throwing out the fictional novels which I had been storing for years to make room for all the non-fiction I was attracted to. I would be walking passed a book store and be drawn inside the shop to browse. Nine times out of ten I would leave the shop half an hour later with a new addition to the book shelf and each of them seemed to answer the questions I had at the time. To date I have a library of at least 150 titles covering my interests in crystals, Homeopathy, fermented foods, the spirit world, Para magnetism, plant spirits, how your hands can heal, devas in the garden, mind power, the link between soil, grass and cancer, Weston.A.Price, Rudolf Steiner, herbal antibiotics, and Biodynamics. While some of them may not be directly related to my profession each one has helped to increase my knowledge and ability to use all my senses and become more aware of what science cannot explain nor needs to.

CHAPTER SEVEN

Stress is a Killer

We lived and worked on the Willow Road property for three years. By that time milk production was beginning to improve and the soil tests were showing positive results (tests included in chapter fourteen). Our organic methods were working extremely well but a member of the owner's family didn't see it that way. She had not seen a fertilizer truck spreading phosphate or solid fertilizers since we'd been there. With the help of a farm advisor we tried to explain our fertilizer programme but it was impossible to make that particular family member understand, the consequences of which our lease was not renewed. We were absolutely gutted. We had managed the property as if it belonged to us and were very proud of our achievements, especially the soil test results, the new electric boundary fences, the $4500 upgrade and fencing of the tanker track, up-grading the house and gardens and putting the property back into dairying. None of this helped our situation, we had been told to leave.

The search was on for another organic dairy farm. As at 31st May the herd still had 3 months before completion of the 3 year term for full USDA certification, so we could re-start organic certification but we just needed to know where. We had been tenant farmers for the past twenty years but that was about to change.

We were offered a 50/50 sharemilking position on River Road. The plan was to increase the herd numbers from 140 to 170 as the owners felt their property was capable of carrying extra stock after converting it to organic over the previous five years. We were to milk thirty of their animals to make up the numbers. My daughter and I were familiar with the property as the owner wanted to lease it out the same season as we shifted to Willow Road. We had a tour of the farm in March of that year and at that stage the property was covered in penny royal, a smelly and inedible herb and a lot of summer weeds. We weren't very impressed and decided the Willow Road property was the better proposition with the owners themselves suggesting that was the best way to go. We became good friends though and often attended the same field days.

Now, three years later we didn't have a choice. There were only a few organic farms around, and none wanting a sharemilker. We even looked into properties to lease with the plan of beginning organic certification from scratch for the third time, but there were none. It was never an option to sell our cows and continue farming with someone else's herd, our girls are part of the family. They are amazing creatures and have given their life's blood for us to survive in the agricultural industry. We owe them a lifetime of gratitude.

On Gypsy Day 2012 we were moving to River Road. The farm had improved tenfold in the three years from the time we first saw it. Three truck and trailer loads of cattle, the furniture truck, the truck carrying the tractor, implements and walk in chiller unit plus numerous car and trailer loads of farm equipment, all moved north. The young stock were trucked up six weeks later because they were grazing on our organic dry stock block we continued to lease in Whatawhata. We hadn't dehorned the rising one and two year olds as I was trying to understand the principles behind biodynamics and I began to wonder if the bovine horn served another purpose besides a defensive role. The biodynamic preparation 500 is cultured cow manure collected from lactating cows. It is placed inside dried cow horns, taken from a lactating animal and buried in the earth over winter. Steiner envisaged the points of the horn reaching towards the heavens and drawing universal energies back into the animal. If this was happening while the horns were still attached to their heads, what natural forces were we interfering with by cutting them off?

Our new employer requested all the horns be removed. We didn't have a problem with that, it was just one of our experiments where some cows were well behaved but others used their horns and were very domineering. The Animal Welfare Act requires any animal older than twelve months must be anesthetised before their horns are removed. As suggested by the biodynamic group, we put castration rings on the horns six weeks before we appointed our vet the job of cutting them off. She noticed the horns lack of blood circulation and decided to cut the first lot without anaesthetic. The castration rings had made a huge dent in the horn and she was very surprised by the small amount of blood loss when the horn was removed. There was also a lot less stress on the animal than when she used anaesthetic. She was utterly amazed and couldn't see any reason to have them anesthetised. After being dehorned the animals went back to the paddock and started eating straight away. They didn't seem to be in any pain. Earlier in the year we'd applied the rings to a couple of animals who were being bossy and we saw the horns were falling off after a period of time with no cutting required! On a couple of them the rings didn't last the distance and broke off, but they still made an indentation in the horn making it easy to apply another one.

In earlier years we dehorned the heifers when they were young calves at six weeks of age using a hot iron, but we noticed their heads were sore for days and they hurt themselves when they were drinking their milk. We also felt they weren't as quiet and didn't trust us the same because their little heads were still burning and they remembered the pain. It was cruel. We now cut off their horns before they are ten months old. A few short stumps occasionally grow back but the ends are rounded and not as dangerous as a mature set of horns. By doing it at that age I've noticed their heads have a better shape, growing a lot squarer as they mature and our new routine is so quick they still trust us after the event. Before letting them out of the dehorning bail they all get a spray of

Arnica (homeopathy) on the nose to stop the bleeding and help with any bruising.

Our move to the River Road property wasn't without its challenges. I made the suggestion we could have a white board to write down and co-ordinate any farm jobs, which wasn't such a great idea. Then we had the highest somatic cell count (white blood cells) in our bulk milk supply that we had experienced anywhere. As our herd calved and came into milk, we took a sample of milk from each individual animal and tested it with the Mas-de-Tec to ensure their milk complied with the standard required to send it to the dairy factory. We had so many not making the grade we began looking for a probable cause. The milking machines got inspected for the second time, we changed the cup size of the clusters to better suit our predominately Friesian cross herd and we added AgriSea tonic and SSC (homeopathy) to their water supply. Slowly the situation improved but the cell count remained the highest we'd ever had to manage. Forty nine of the 170 cows in the herd had mastitis infections that season. Just about thirty percent of the herd. Far too many! Some mornings we'd have two or three showing signs of mastitis, then things would calm down for a week before another lot would come in. They seemed to come in waves. The owners had had trouble the year before so something was out of balance, maybe the pasture or soil or was it simply a stress factor?

Another challenge for us was finding the reason behind the herd's smelly and runny cow pats. We thought the situation would improve when the girls had some hay and silage to eat but there was no change. The cow pats stayed the same for the whole season. To me a sign of good nutrition is when a cow pat doesn't splatter everywhere but instead makes a *perfect Pavlova* shape when it falls to the ground. We were not seeing this at all. Just because the farm was certified organic didn't mean it was perfect by any stretch of the imagination. Some of the herd also grew extra *skin* on their teats, it was like a soft stringy wart. We had never had them before either. Something was definitely wrong.

The blood tests taken from the cows on our arrival showed a reasonable mineral balance but the vet singled out the low selenium in a few of them. He suggested we would need veterinarian intervention to get the levels up otherwise we would have retained membrane problems after calving and trouble getting the herd back in calf. Because the vet couldn't supply any solution allowed to an organic farmer, nothing was done. The approaching calving season was going to be a short one because the girls got pregnant very easy on the Willow Road property. Within four weeks 160 of the 170 cows had calved but not one of them retained the placenta after calving as was suggested. We had no problems getting them back in calf either and only lost one cow due to a misadventure when she went to sleep on uneven ground and couldn't get herself out. The vet's intervention was not required! Were all the tests wrong or did the homeopathic selenium that we added to the cows water supply remedy the situation? I've noticed the vets are wanting to play a larger role in how farmers care for their animals and manage their land and the push to use their products has increased tenfold.

Our herd's management wasn't our own on River Road as the owner insisted we follow the instructions given. I did for the first six weeks when all the winter grazing had been mapped out ahead of time giving us a chance to experience how the paddocks *ran* together but then I was asked to graze thirty late calving cows in the steep country. I didn't like the idea. We didn't have any and I wasn't going to put heavily pregnant cows on steep country ten days before they calved, so I left them on the flatter paddocks. My plan wasn't received all that well.

The owner also took exception to us erecting a temporary tape around the head of a culvert which we saw as a hazard to our stock. We wanted the tape there as a visual barrier for them not to walk too close. The tape was taken down on several occasions which was difficult to understand. It in no way restricted a vehicle or a person from passing the area. On one occasion we forgot to hang the *directional tape* back on to the fence line and the message was highlighted and conveyed to us via the white board. The tapes were used to guide the stock into each individual paddock so they wouldn't walk beyond the gateways.

Cleaning out the sand trap at the dairy shed was one of many medial tasks we were instructed to do via the white board in the lock up room. We thought the board would be a good idea to write down our farm jobs. It was not. Cleaning the sand trap was a weekly ritual we had no problem with but to read it on that board every time we turned around was instant pressure and it was detrimental to our working relationship with the owner.

A few days after our arrival I made a batch of EM so we could use it as a teat spray on the cows after each milking as well as a fertilizer. Our main concern was the farm's bloat history so we used the four wheel motorbike to spray seawater, EM and fish fertilizer on the paddocks at least a couple of times ahead of the bloat season. We were very conscious of the owners telling us they had lost ten cows the season before and I knew my mixture would help, especially the seawater. There is nothing worse than watching your pride and joy dying from too much gas in their gut. The cow literally chokes to death as the pressure of their enlarged stomach becomes too much. Their natural instinct is to stand with their front feet on a rise so they can breathe but it's a horrible way for them to die and we had to do all we could to prevent it. Being organic I'd learned fish oil was a possible alternative to chemical drenches but there wasn't much else.

The landowner was very enthusiastic about the seawater idea and decided to take a trip to the ocean and get enough to do twelve paddocks at twenty litres per paddock. That was 240 litres of which their larger vehicle could easily carry. We were overwhelmed because it was a two and a half hour round trip to the nearest beach and would have meant two trips for us. We waited in anticipation. The fresh seawater had to be applied to the pasture within three days of being collected before it lost its vibrational energy and all the microscopic organisms died. We were very disappointed when no seawater arrived and the landowner's proposed plan didn't eventuate. We wanted it sprayed on the grass to help prevent the cows dying of bloat. We managed to travel to the open sea and collect a hundred litres a few of weeks later.

The bloat season arrived and paddock number seven was full of fresh clover and looked as if it could give us trouble. We fenced it for two nights grazing and fed out a couple of haylage bales in the first half. After we'd finished milking that night my daughter and I went to the paddock to inspect the girls and despite the haylage we'd given them saw a large number of cows starting to gas up. We walked amongst the girls for what seemed a life time keeping an eye on the worst ones. I hated it. We walked a couple of them to the shed and gave them homeopathy which seemed to work and by the time we returned them to the paddock the others had degassed and the worst was over. The following day we sprayed some seawater on the other half of the paddock and fed out two more bales of haylage before the herd went in. My plan was to alleviate the bloat by spraying seawater on the grass closer to when it was grazed. To our relief there wasn't a sign of bloat, the seawater had worked its magic! Suddenly our employer instructed us to stop all seawater applications plus any other fertilizers we had made. This was very difficult to understand. We were in the middle of the bloat season and the owner had taken away our only form of prevention. The next couple of weeks were extremely stressful hoping and praying the girls would be okay as we watched and waited for the danger period to pass. Thankfully luck was on our side and we got through without any casualties.

Around that time we had another challenge to face, it was the invasion of the ticks. We had never seen so many on any one animal before! At each milking my daughter and I pulled hundreds of big ticks off their udders and stomachs. As we collected a handful we put them in heaps out on the concrete for the blackbirds to eat. We felt so sorry for the girls, we had to find something to help them get rid of the ugly blood suckers. Amazingly we had already been working with the answer. We had added BioSea fish oil to the teat spray as a skin conditioner and when I noticed the ticks in our heap were crawling away, I sprayed some teat spray around them. They froze in their tracks, they didn't like it at all. From that day on we smothered the udders of the worst affected cows with BioSea and noted their udders remained tick free for several days. We even spread the fish oil along the backs of some cows hoping it would get into their system and deter the ticks. The young stock also suffered as their ears began to curl up on the ends with all the smaller ticks sucking on them. Fortunate for us our baby girls were really quiet and enjoyed us giving them a rub so we would take a bottle of BioSea with us and rub it on their ears and udders while we stood amongst them in the paddock. They loved it and the worst affected animals were pushing themselves forward to get our attention and their share of the oil.

At our October shed meeting with the owner we got an explanation to why we had to stop the seawater.

We were instructed to check the organic rules around applying fish fertilizer and seawater on the pasture the day before it was grazed with the milking herd. The fish mixture was perceived as a problem because a farmer's stock died after grazing a treated paddock and the owner didn't want any of his cows dying. I assured him we didn't want any dead cows either and the fertilizer was safe as long as we added EM to neutralise the *bad bugs*. BioGro gave us their consent but the owner refused to allow any seawater or our fertilizers on their property. Not long after our meeting, my remaining fish fertiliser drums were emptied out on to the tanker track without any explanation. It was one stinky mess. At that stage we had been using the fish fertilizer and EM on our dry stock block so it was very disappointing to lose it like we did.

A few strange things began to happen soon after. On three separate occasions items belonging to the farm disappeared from the lockup room and I was worried someone had come in from the road and taken them. A short time later a couple of test buckets also went missing. I asked the owner if he had seen anyone suspicious around the dairy shed. It was strange the test buckets had disappeared during daylight hours. We had so many cows coming in with mastitis we were using the extra test buckets belonging to the owner, to keep the infected milk out of the main supply. I was lost for words when the owner's garden shed was unlocked and I was handed the buckets in question. I was really worried about what else might be removed from the dairy shed and what was going to be written on the white board next. We were on edge and I couldn't take much more so I decided to lock the room after every milking for my own piece of mind.

I was able to disconnect myself from the farm every Friday when I drove to Thames to partake in the level three Organic Horticulture course. I really appreciated the outlet but when I returned home, nine times out of ten I would hear about an awkward situation that occurred in my absence. On one occasion it involved the owner spreading slimy, smelly sludge on pasture that was only a couple of days from being grazed with the milking herd. It was at the time we were fighting mastitis, runny cow pats and high somatic cell counts and we were looking at every possible cause. Our daughter asked if the sludge had been tested to which she was ignored and the process continued. She was extremely upset and by the time I got home there was nothing I could do, after all, it wasn't our property. The girls didn't like the contaminated grass and we had several new mastitis cases after they grazed it.

I was fortunate to have my organic course as a distraction to the stress but the best part was returning home to Ethan who was growing into a wonderful, happy and healthy three year old. He was our resident clown who never failed to keep us laughing and every day he smothered us with his charm and good *medicine*.

Early November the owner emailed us the date of the next shed meeting and informed us an organic representative would be taking us through our agreement to clarify a few things on that day. I don't recall the articles that had to be clarified but I do remember they also wanted to discuss their change in farming policy. We had no idea what the change in farming policy meant. The organic representative took four hours to go through the sharemilking agreement from an organic perspective and we agreed to all the amendments. We signed the extra clauses and added them to our agreement. Finally it was the last item of the day, the change in farming policy. Then the bombshell was dropped. The owner was not going to continue organic certification beyond that first season. The farm was in its final season of a six year contract with Fonterra and the rules were putting too much pressure on the organic farmer. The turn of events that day was unbelievable. We had spent four hours ratifying any loop holes in our four year 50/50 agreement and it was all for no reason. 'Was this a joke?'

It was no joke, in fact the joke was on us, we had signed a four year agreement but it wasn't an organic one. The word organic was missing from the title on the front page. But the owner knew of our passion because we'd been to so many functions together and shared so many organic techniques over the years. The initial offer when we signed our agreement was for us to supply organic milk to Fonterra for the last season of their six year contract and then we would supply organic milk for another three years. It obviously wasn't going to happen.

The owner gave permission for us to move and if we didn't find another position, we could stay. We began our search for another certified organic property. Meanwhile our employer was making plans to feed our girls molasses in the following season but our farming practices had changed so much, I knew I didn't want a bar of it. I didn't want our cows eating anything other than green grass. Molasses had been used on the property in the past and all the equipment was still there but the molasses had left holes in the floor of the shed. I couldn't understand why our girls had to have the sugar that corrodes concrete and rots teeth.

In December the organic certifier gave the landowner permission for roundup to be sprayed around the river banks as long as the residual was at least half a metre away from where the animals were grazing. The rule of compliance didn't work and we had to put up electric tapes around the fence lines where the spray had spread out on to the pasture. There was also some convolvulus creeper sprayed along the cow lane. We realized what had happened when we saw its leaves wilting and I was really concerned it would jeopardise all of Fonterra's organic milk supply if the cows ate it. We could've been prosecuted big time! The milking cows walked by the area on the way to the shed and loved to eat the creeper growing under the fence. It was also a drought they were eating everything and anything they could find.

At the end of January the owner of the property informed us we would be supervised by a farm consultant for the rest of the season because he was going on a three month holiday around New Zealand. Once a month I sent in a report which was forwarded to our employer after the consultant walked over and inspected the property with us. The consultant was also instructed to collect the owner's share of the cash we routinely paid from selling some raw organic milk. It was around $150 per month and I couldn't understand how a complete stranger was more trustworthy than us?

As if we didn't have enough on our plate, it was a monstrous drought that summer. In January, February, March and April we got very little rain. At the beginning of December we were on a thirty five day milking rotation of the paddocks and by the January 25th we had extended that to forty three days but even that didn't see us milk beyond the end of March. The drought was taking its toll and the supplements were flying out the window. We began feeding supplements on January 31st and we had 230 bales of silage and sixty two large hay bales in storage. I thought we had plenty but we had to bring extra feed up from our organic grazing block in Whatawhata. Thank goodness it was still ours to lease because it's not easy to find certified supplements and being organic, we were only permitted to use certified organic inputs. Our consultant did the feed budget and concluded the property fell short of growing adequate pasture for the coming months and said the owner would pay their share of the extra haylage bales bought in to feed the stock.

Meanwhile, we had advertised ourselves on the organic website in search of another position. In doing so we came across an organic dry stock block for sale. It was a property that could easily be put back into dairying and if we could find an equity partner we would be in with a shot. Easier said than done. The vendor said he would take it off the market on April 1st if no one had shown an interest by then.

The days seemed to go so slow. We couldn't plan either way, staying on the River Road property or moving to Pirongia. The stress was unbearable.

The cows were all dry at the end of March and apart from feeding the animals their silage there wasn't a lot to do so I spent my time trying to find how to alleviate the stress. I was drawn to a book entitled The Book of Natural Devas which was a real inspiration. I always had an inner feeling of some other world removed from our own. As I searched for help the world of natural devas was revealed and I decided I wanted to know more. One particular day I cruised down the farm looking for devas, and out of the blue I decided to *hug a tree*. The idea just came to me. Crazy I thought, but why not. I stopped the bike in front of an old Puriri tree that looked like it could do with a cuddle. I jumped the fence and wrapped my arms around its trunk pressing my entire body against the bark. I closed my eyes, visualising the tree as my peer and trail blazer, and I begged for something to relieve the stress. Without warning my heart felt as if it was going to jump out of my body! It began to throb so strong and my head felt light and sleepy, as if it wasn't part of me. 'What was happening?' I retreated so fast, I didn't stop to think, I was getting out of there! I jumped the fence and took off. I'm not sure how to explain my strange experience and I've been too scared to try and repeat it. I secretly wish I'd continued to embrace the moment and learned more, but I was so shocked to feel an actual response. I later learned from my book of devas I was supposed to ask permission before I entered the tree's space. Maybe that was where I made the mistake but were the sensations I felt an actual response from the tree itself?

Another book that helped pass the time was about Asian medicine where they believe fear and stress are held and recorded within the water element. Our bodies are 70% water and our brain is 80%. If any unresolved emotion such as fear and stress are held within the body for longer than two years it will create a contraction of the body's natural ability to be true to nature. Dr Masaru Emoto, a Japanese scientist, confirmed this reasoning by writing *love* on one test tube of water and *hate* on the second. He swore and cursed at the water marked with *hate* and read loving poetry to the other. He then froze the solutions and photographed the crystals. The *love* solution produced crystals of amazing beauty and the water labelled with *hate* produced chaotic and deformed images suggesting love breeds love and hate breeds more hate.

Our desire for a place of our own on Mount Pirongia wasn't enough to be able to make the purchase so we weren't sure where we were going or if we were moving anywhere. It became increasingly difficult to find enough grass for the cows as each portion of land was taken out of organic certification and we didn't want to lose our herd's certification. There was no rain on the horizon and the girls were rapidly losing weight.

On April 4th the Pirongia vendors rang telling us they had not found a buyer for their farm and wondered if we would like to share milk with them on the property. Yes! The jubilation and the relief was enormous. At last things were moving and we could start to make plans. We still had a stressful eight weeks ahead of us packing up, cleaning the house and arranging all the transport but at least we could move away from the hell we were in. As with our last two shifts, we were milking a few cows over the winter months so the Pirongia milking shed had to be ready to use by June 1st. We travelled to Pirongia several times to help arrange the people and businesses we knew who could do the job.

We informed our employer of our impending departure on June 1st and the farm consultant handed us a job list, mentioning a broken post here and a missing staple there, nothing of consequence, and not to forget the ragwort we had missed. Yep, one ragwort!

We cleaned out our belongings from the lock up room, scrubbed it from top to bottom, shut the door and left a key in the lock. Eleven months earlier I had new keys cut for it because the old one had broken and in case it happened again, I had two cut. That day something told me not to leave both of them in the door. The following morning the owner did the customary farm walk and shed inspection, locked the door and took away the key. A couple of days later I couldn't find our packet of milk filters. I searched for them everywhere without success. We were milking six cows so Ethan could have his fresh raw milk and I wanted it to go through the filter sock before it went in the billy. As I had only returned one of the keys I used the spare one to look in the lockup room. Sure enough there they were, stashed away on the shelf but I couldn't understand why they were there. I locked the door again and contacted the owner. Next day there were four filter socks on the bench in the shed for us to use. A few days later the dog's plastic milk container disappeared. That was strange, it must have been windy and blown away. We looked everywhere, but hang on a minute, I wonder. Again I unlocked the door and there it was. I left the container there because it really wasn't important, we could find another dog bowl. We did, one of the old plastic containers we had had staples in. It too *went walking* and was sighted in the lockup room! The whole situation was becoming unbearable so I rang the trucking firm to see if we could bring the shifting date forward. I moved it to May 29th, three days earlier than the original plan.

A truck and trailer load of our skinniest cows were taken south to Pirongia on May 21st taking the pressure off the available grass. I was conscious of the supplement we had to leave on the property as per the agreement and the grass wasn't growing fast enough. We had skinny cows, very little grass and we were frantically packing and cleaning, and then the unexpected happened. Ten days before Gypsy day hubby had a heart attack. He had had chest pains three mornings in a row and on the fourth morning I drove him to the medical centre not aware of the verdict. The ambulance was called and he was taken straight to hospital. The doctor told him heart attacks commonly occur first thing in the morning and one of his chest pains was a heart attack.

I couldn't dwell on the matter. He was in the best place and there was so much we had to do. I informed the owner of the new departure date and the email arrived 'Okay, we will be over to inspect the house at five o'clock on the 29th, so make sure you are out by then'. The response was unbelievably blunt!

True to the landowner's word, the house inspection commenced at five o'clock on the dot! I had actually forgotten to look at the time, we were too busy cleaning and ensuring everything was spotless and in order before we left. I even remembered to leave the extra key for the lock up room on the kitchen bench. It had served its purpose. I got a huge shock when I finally got into the car, backing it round to leave and saw the owner right there in my rear vision mirror!

I consider myself to be fair and tolerant and I make every effort to get along with people but I was the most stressed I'd ever been on the River Road property. As well as hubby's heart attack, my daughter and I lost a lot of weight and it was difficult to sleep at nights. Even our Labradors had many sleepless nights guarding the house perimeter and barking at every little disturbance, everyone was on edge.

Six weeks after our arrival in Pirongia, I realised just how much stress my body was under. I woke one morning and my nose was completely blocked up. It was really odd. I blew my nose into a tissue and the snot, pus, call it what you like filled the tissue. I just blew and blew until I could breathe again. It was disgusting. I had to do the same every morning for six days! I didn't have a cold or the flu and I didn't feel sick. My blocked nose only happened while I was sleeping. I can only conclude I was so *uptight* and so stressed that my body was finally able to relax and release some pressure.

Four months later I had to attend the Arbitration court with the River Road property owner. There was a disagreement over whether our stock had enough to eat on the farm without bringing in more supplement during the drought. We were fortunate to have the consultant as a third party and the matter was settled with the Arbitrator ruling in our favour to the value of eleven thousand dollars.

CHAPTER EIGHT

The Spiritual Mountain

Gypsy Day 2013 and we were moving back to the Te Awamutu area. We had always wanted to live up Mount Pirongia with its beautiful climate and volcanic soil and where they never had to worry about the summer droughts. We signed a one year organic 50/50 contract with a fantastic family. They were great to work with and shared our organic passion. The plan was for our contract to be rolled over providing everything went okay. There was a farm consultant to bounce ideas off and we all worked really well together. When we arrived there was plenty of grass for the cows and the milking shed had been upgraded and was ready to go.

To me it was and always will be a very special and spiritual place. The property itself had beautiful freshwater streams running down the gullies and was the second to last farm before the bush line of Mount Pirongia. The view was to die for, absolute magic. After dark you could see the Hamilton city lights blazing down below and to watch the lightning of an electrical storm play around the hills in the distance was an incredible sight. For the mountain itself, I was grateful to be watched over by such a majestic icon looming high above us.

The day of our move from the River Road property, I became quite emotional as I drove into the Pirongia settlement. I think it was both a relief as well as a feeling of returning home. Everything felt right. I had a good cry as I drove the car towards our new home. It felt like I had been in prison and I had regained my freedom.

The Pirongia aura revived us all, alleviating the stress of the few weeks before. Hubby was in hospital for a week where he had a stent put into one of his arteries and then stayed with our eldest daughter while we moved house. Physically and emotionally we were all worn out. For me I was privileged to find the ultimate vantage point to relax and watch the fresh mountain stream, bubble and dance between the volcanic rocks on its way to the river below. It was magical. At times I felt like an intruder and wondered if I should have asked a higher power for permission before entering the area, it was so spiritual, so untouched and so beautiful.

Despite the serenity of our new home I was still looking for answers as to why I was feeling so *disconnected*. I made an appointment with Marnie McDermott a lifestyle coach who I hoped would give me some guidance. I explained I felt as if I couldn't go any further, I lacked direction. It felt like there was some sort of blockage, a wall in front of me. I was rather sceptical at first as I hadn't tried anything like it before, but I was willing to keep an open mind. Marnie explained the format as I laid down and started to relax. I was then asked to say the affirmation 'I am at one with myself and all my creation'. I couldn't say it, the words wouldn't come out of my mouth no matter how much I tried! She placed the crystals on my chakras as I listened to how my soul had returned to the earth fifty two times but at that moment my soul was in five pieces, it wasn't whole. The time of my soul's demise was back in the twelfth century when I was born as the illegitimate daughter of an American Indian chief. I wasn't recognised as the chief's daughter and was raised by the medicine man. I grew up with a deep understanding of the earth, the animals and the plant kingdom, so much so, the medicine man said I could take over his role in the tribe. The Chief wouldn't allow it but said, if I survived the twenty one days of initiation, he would consider the request. So I was sent into the desert without food, water, or weapons for twenty one days. On the twentieth day of my quest I was crouched in prayer when my heart was pierced with an arrow. My soul was split into five pieces. At that stage of the story I was crying and the uncontrollable tears were streaming down my face. I didn't know why and I couldn't stop them. As I was being spoken to I experienced *an overwhelming healing* and I heard my soul saying 'Thank you, oh thank you'. It was an awesome experience. The test then was for me to say the affirmation that I couldn't manage earlier. 'I am at one with myself and all my creation'. I was surprised to hear the words come out, albeit a bit shaky. I had just experienced two hours of the most uplifting and amazing sequence of events. I felt as if I was reborn.

I wanted to share my healing experience because I

believe it allowed me to turn the next page in my life. After the experience I felt more grounded. I was more aware of *me* and especially aware of my empowering soul that I'd discovered a connection to. I returned home to Mount Pirongia as a new person.

When we shifted to Pirongia, I had three weeks of my Organic Horticulture course to complete in Thames. It was now an hour and a half drive to get there, but it was worth it. I really relished being with such an amazing, likeminded group of people and while I entered the class feeling rather drained and vulnerable, by the end of the day my batteries were recharged and ready to go.

We increased our number of raw milk customers, some of whom had travelled long distances, firstly to the River Road farm and now to Pirongia to collect our cow's milk. They all commented on the incredible view we had from our new place. Some of them could feel the spirit of the property and commented on how special it was. One woman familiar with the biodynamic concepts revealed *the birds know* and my intuition told me she was right. We were privileged to have kiwis calling out in the bush right outside the house and many other birds seem to come closer to us than you'd expect. I was sitting beside a tree one morning and a Grey Warbler came out on a branch beside me. The Grey Warbler has been a lifetime favourite of mine because of its stunning bird call. It was fantastic, we were all smiling again. The cows were happy, we were happy and with our new employer we made the perfect partnership.

We had a beautiful spring that year comprising of

four weeks without rain or the usual stormy weather. Of the 150 calves born we only had one death. It was extraordinarily warm and dry, exactly what we wish for every calving season. The only real problem we had was the number of cows with sore feet. At one stage there were thirty three so I decided to keep them as a separate mob to give their feet a chance to heal. I wondered if their hooves were too soft and they were showing a zinc deficiency. Every place is different and it was unusual for us to have so many animals with sore feet.

On the River Road farm I recorded every activity on a daily basis, even the rainfall but in Pirongia it wasn't required. At last we were feeling appreciated for doing our share of the work. We had a one year organic contract which would be rolled over for the next two seasons. That suited us fine. Our two best employment positions were based on annual contracts and were both rolled over for seven and eight years respectively. Our farm consultant reiterated his faith in the landowners keeping their end of the bargain. We were tired of moving and simply wanted to take care of our girls and continue farming under organic certification.

The property owner spent a lot of time and money getting everything ready for the new season with a new outside vat stand, a concrete tanker pad and extended the milking platform from a twelve aside to fifteen. We installed our stainless steel cups, cooling system and ultra violet light in the dairy shed. My daughter and I planted out the Tagasaste (tree Lucerne) that I had grown for my organic course, several fruit trees and at least fifty native flax plants we had split from those on our Whatawhata dry stock property. We had found an authentic compatible partnership at last.

We introduced the owners to the EM we were making which was applied to the property along with a liquid fertilizer containing humates. Many of the gorse gullies had been fenced off and were being returned to native bush as a controlled area. A particular nasty pasture problem appeared as the dry months arrived and that was yellow bristle grass, the unpalatable invasive grass we'd seen on Willow Road a couple of years earlier. The district had huge problems and farmers were getting their heads together to find a solution. The only one available was mowing it down to ground level to try and prevent it from spreading because there was no weed killer to stop it. Yellow Bristle grass first appeared where the councils had sprayed Roundup around the roadside marker pegs and was now invading pastures on many properties around Pirongia. Since then I have seen it growing in many other areas, even in Northland.

All the family came to our place for Christmas that year which was very special because our eldest daughter was going to live in Canada shortly after. It was the last opportunity to have our four children, their partners plus our nine grandchildren and extended family all together. A few months later I would come to realise how valuable our Christmas gathering was. In early February our eldest daughter left a few of her favourite items with us and left New Zealand.

For the New Year we caught up with friends in Te

Awamutu, something we hadn't done for a couple of years. We were starting to relax and catch up with the things we wanted to do. Two weeks later the landowners arrived at the house and I thought it was a little strange because they weren't dressed in their usual farming clothes. We invited them into the house and sat at the kitchen table. As he wasn't one for small talk, the real reason for their visit was soon revealed. They informed us they could not renew our sharemilking contract. An empty, confused feeling came over me. 'What was wrong? What had we done?' As we sat there I had my head down to hide my tears. They explained their son wanted to return to the farm. We were lost for words, dumbfounded and disillusioned. The conversation was a bit strained and awkward, but it was okay, we understood. I began to wonder what was happening. It felt like we were stuck in Groundhog Day, acting out the same story over and over again. Surely it wasn't our fault, we were a hardworking and conscientious family, and we were good caretakers of the land but here we were again being moved on to make a home somewhere else. Why was the universe doing this to us?

We had five months to find another home, so there was ample time to get organised. Our employer said we should get our own property. He said we were more than capable and we should be our own boss. He had come from Northland and suggested looking in that direction where land was cheaper than the Waikato and Taranaki regions.

My immediate thought was we would lose our organic

certification and all that we had worked so hard to keep, but then again it would be our own place. The options were there: find our own place, 50/50 sharemilking or an equity partnership, somewhere. The search began, again.

We started by ringing friends and family to see if they knew of anywhere and I put word out on the ODPG, Organic Dairy Producers Group website. The Farm Employees Wanted section in the newspaper was unusually quiet that year, farmers weren't moving around as much. I advertised on Fonterra's webpage, with no response. Our herd of 170 cows was too small. A couple of contract milking positions came up but that meant selling our girls or leasing them out and we were not keen on either of those ideas. We had years of breeding behind our herd and they were family. We also didn't like the idea of adding supplements such as palm kernel to our girl's diet and farm management. It was foreign to us and our cows.

We pretty much searched all the Waikato and Taranaki regions without success. At the beginning of February I received an email from an organic farmer in Northland. He was looking for a sharemilker and possibly an investor. Finally we had a lead, something to hope for, but it was a long way from the areas we were used to.

We weren't sure of moving to Northland, away from family and friends so we continued advertising in the local newspapers. Nothing was happening.

I mentioned our situation to the farmer grazing our weaner calves. He was a Botanical Homeopath Practitioner and we really appreciated the advice he gave us about herbal remedies and the like. One day he mentioned his bad experience taking on stock that came from a property where chicken manure had been spread on the land. I was aware the antibiotics and GE in chicken food could enter the food chain this way disrupting an animal's metabolism but he actually witnessed the animal's purging for several months as they tried to get rid of the toxins. He said he watched his land deteriorate because of their manure, growing more weeds and thistles and would never have them on his land again. His experiences were invaluable. When I told him we were thinking of moving north he said the position was not the one for us. He said 'I can't see the full picture, I can only see the headlines and it is not the one'. I was moved by his concern and respectfully told him 'We may not have a choice'. I quietly hoped he was wrong.

Meanwhile, the summer drought had set in. So much for the Pirongia Mountain not being affected by drought conditions! Our farm consultant put us on once a day milking on January 20th and insisted we move the herd to a fresh paddock each night. To this day I do not understand his advice because we gained nothing. Normally we change our milking routine to conserve pasture. That is, milking once a day meant one paddock for the whole twenty four hour period until the next milking. He was adamant the extra paddock given to the girls would hold their milk weights and keep their condition on, but neither eventuated.

The herd was dry by March 25th and we had to send thirty of them out grazing on an organic property in

Piopio in the King Country. Another fifteen joined them three weeks later as the organic supplements were running low and we had to keep in mind our obligation to leave the correct amount for the following season.

Once the herd was dry we split them into groups of three and four per paddock for a couple of weeks to conserve the silage while we waited for the autumn rain to come. It worked well, but it was about then the yellow bristle grass began to grow and take over any bare ground. Any hint of moisture would encourage its growth. When I see an invasive plant I automatically question why it wants to grow where it does. What is in the soil that allows it to flourish or maybe what isn't in the soil making it the perfect home? Our immediate reaction is to *kill* it but finding some answers makes more sense in the long term.

On the home front we were packing things into boxes but with nowhere to go. It looked like Northland was the place but the property owner hadn't specifically agreed to take us on.

We decided to look at the Northland properties that were up for sale. We went with a friend and bank manager to look at one north of Dargaville. Huge house, rolling contour, a lot of kikuyu grass and yes we could make it work. Unfortunately our financier said no, so that was the end of that idea and now time was running out. It wasn't that we couldn't service the loan, it was our lack of equity.

As we were up in Northland we arranged to meet the farmer who emailed us in February and have a look at the position he was offering. First impression was the whole place was in need of some TLC. The main house was smothered in creeper and large privet trees and the dairy shed had a lot of old rubbish around it. We also noted the enormous brown area around a diesel tank that was seeping fuel. Obviously the diesel had been leaking for some time. I wondered why nothing had been done about the huge diesel stain. There was gorse covering an enormous area of good land and being organic we saw it as a huge challenge to get it under control without the use of conventional sprays, but the land had potential.

We travelled back to the Waikato with positive vibes. The area had had the rain that summer preventing the drought conditions we were seeing in the Waikato. All we wanted now was for the owner to agree to take us on.

At the beginning of April the Northland property owners travelled down to Pirongia to see our herd and where we lived. He was more than happy with the cows but enquired if we had tested them for A2A2 milk. We hadn't, but the majority of the Livestock Improvement bulls we had used in the past favoured A2A2 progeny. The majority of cows produce a mixture of A1 and A2 beta casein. They are different proteins with A2A2 said to be more easily digestible. The cost of testing for A2A2 milk isn't cheap and not just a one off because as new heifers joined the herd each year they would have to be tested as well. Surely this wasn't going to be the deciding factor as to whether we got the position or not?

At the dining table they began to fill in the facts surrounding their finances. They would offer us a 60/40 organic sharemilking position, 60% of the milk production for them and 40% for us. They felt they couldn't afford the traditional 50/50 agreement. Well, this was different!

Initially we said a definite no. At the time our daughter didn't feel comfortable with the idea. It was hard enough surviving where we got 50% of the income, and we were being asked to accept only 40%, and Northland organic suppliers weren't getting paid the premium for their certified milk like the Waikato.

I think we all felt the awkwardness but we denied our feelings because we were getting desperate. It was the middle of April and despite all our phoning and advertising not one farming position was available for us. Plenty of people were saying 'Don't go north, there's nothing good about Northland' but where was the alternative?

We went through all the pros and cons of moving so far away from our family and friends and finally decided the offer was do able. It was going to be tough but we worked out the lower percentage was equal to the lease position we had been on a couple of years before. We also concluded we would accept the lower percentage but for one season only after which it had to be a 50/50 split.

I recall the landowner's decision wasn't immediate and it was May 12th before we travelled back to Northland and the agreement was finally signed. A2A2 milk wasn't a condition to us filling the position but he explained the original herd on the property was tested so their raw milk could be sold with the identity.

It had been a long drawn out affair but finally the

agreement was in black and white and we had a new property to move to. We then had to organise trucks, trailers and furniture movers for the upcoming Gypsy Day. It wasn't like previous shifts where we took a few trailer loads to the new farm in our spare time. We had to travel more than five hours to get to our destination. Exciting, but a huge challenge as everything had to be shifted on the same day.

Thankfully we knew who to turn to at such short notice and one was Road Haulage, the trucking firm who we called on for the third time in three years to shift the herd. It meant a six to seven hour journey for the cows this time. When I booked them in there was a surprised 'What? Again?' on the other end of the phone, to which I replied 'Yes again', and my explanation followed.

We had to co-ordinate forty five cows to be collected from Piopio, one hour south of Pirongia, forty five rising two year old heifers from our dry stock block at Whatawhata and ninety cows from the Pirongia property. Believe it or not, the number of animals at each property was equivalent to a truck and trailer load of animals, with the ninety cows on the home farm filling two truck and trailer units so that part worked in extremely well. Road Haulage could fit us in on May 29th and yes, the Northland landowner gave us the okay.

Then there was the furniture. I rang four furniture movers without any luck, they were all fully booked. I contacted the fifth business on my list and finally had some success. I booked them in for May 29th and hoped like hell their truck was big enough. I said we were moving a five bedroom household, which we were with our eldest daughters extra furniture as well as a sewing cabinet, office desk, three queen beds plus four singles. Not much I could do about it now the date was set and I didn't rate my chances of finding another available truck. It would all fit in, it had to.

Next on my list was a shipping container to pack all the farm equipment in; two quad bikes, effluent irrigator and hoses, milk chilling unit, dog kennels plus more. Because we had a twenty year history of leasing properties we had virtually the same list of equipment as anyone owning land. We hired the biggest container available and paid $2,400 so we could have it delivered for us to pack. Everything had to go into the container in order of importance. This meant the two quad bikes had to be last so we could continue using them right till the last day. We also had to consider the container was going north via the railway and would probably sway around a bit on its journey. We managed to slot everything in to secure the contents from any movement and it was surprisingly full when it was collected and transported to the railway on May 28th.

The tractor, feed out unit, large tractor trailer and the walk in chiller also had to travel north. So that was another truck and trailer unit but it had to be equipped with a Hiab to manoeuvre and load the chiller. $2,500 was paid to the carrying company and everything was loaded and sent north the same day as the herd and furniture. It was due to arrive in Northland on May 30th, the following day. We just hoped the outside motor on the chiller would get under the low bridges through Auckland. The driver measured its height and said there was centimetres in it.

We had become expert packers and co-ordinators by this stage. We had had plenty of practise! As we packed up the house all boxes were marked with what was inside them and we always started emptying the top cupboards and worked our way down. We hadn't had time to hoard extra stuff and as Gypsy Day drew closer anything we didn't want was given to the Salvation Army or sold on Trade Me. There was no use carting it from one place to the next if the item hadn't been used, it was only taking up valuable space.

Gypsy Day 2014 came around and we said our goodbyes to the Waikato. It wasn't the first time we'd moved away from the mighty Waikato, but this time we were moving beyond the Auckland region, through the city plus a further three hours travelling north. At least we'd been able to make day trips to celebrate family birthdays or special events but now that wasn't possible.

Hubby travelled up with the cattle trucks and my daughter and grandson followed me in their car while I was driving the Station Wagon towing the trailer. In the trailer were several pot plants and the outside furniture and in the car I had our three cats in their cages and the two Labradors tied down in the back seat. Two of the cats were really vocal for most of the five and a half hour journey but the dogs were extremely well behaved, curled up beside the cat cages fast asleep. It was a huge disappointment to find all our furniture didn't fit into the truck no matter how hard they tried. Bugger! So three weeks later we had to travel back for another car and trailer load and collect our six chooks.

To release the stress and help me gather my thoughts during the big shift, I stepped back from the mayhem and recorded my emotions using pen and paper with the following capturing one of those special moments.

Leaving Home

The cool presence of frost outside
The glow of hot coals on the fire
Sunrise and my dreams fade.

Cat curled by the hearth
Feeling the warmth of the fire
Her peace symbolic of our homeward journey.

Home is where the heart is,
A dream to be completed
A dream to be fulfilled.

Aye, what a beautiful morning
Father sun is on the horizon
And mother earth beneath my feet
The mountain aloft waving us goodbye

As we prepare to load our girls.
They won't be left behind,
They are family too

The girls stand patiently waiting
Glowing spotlights beckon them forward
Working together, man and beast.
Doors close, hammers down
Up and away on the crated trucks
Their new home awaits.

To move so far from all we know
A challenge for our souls.
Our lives have been so close
But maybe not entwined

Distance draws no lines
As we walk different paths
Not knowing what future holds,
I embrace this day as one.

I drove down the tanker track on the Northland property around seven o'clock that evening. Hubby had rung via the mobile to tell us the cows had made the trip without any dramas, in fact they walked off the trucks looking as if they had just travelled around the corner. Wow, that was a relief. The other bit of news wasn't really what I wanted to hear. The house we were to move into still had people living in it. The landowner had organised for us to stay in a holiday venue just around the road. I was too tired to be jubilant, if that was what I was supposed to be. The day had begun long before day break that morning and there I was fourteen hours later, no closer to my bed.

My first thoughts were for the animals I had in the car. We couldn't waltz into an up market holiday unit with two dogs, three cats and wearing gumboots. But we had no choice. We decided to leave the Labradors tied up at the cowshed on the farm but we would have to take the cats with us. The unit was fantastic, situated in a quiet part of town and separate from the other dwellings in the complex. We let the cats out of their cages, set up their dirt box and hit the pillow. Despite our exhaustion sleep was a little hard to find that night because we were wondering why there were still people in the house on the farm. May 29th was confirmed as the day for us to move in so why was the house still occupied? Our furniture was due to arrive at nine o'clock the following morning and we were wondering where we were going to put it.

We were awake early that next morning concerned about the cows and the new farm. When the cows came off the trucks the day before it was semi dark and hubby had difficulty finding any grass for them. All the paddocks he'd seen looked the same with very little cover. 'There must be more at the back of the property. It was too dark to see.' Day break showed us the whole picture. There was no other grass for the girls and the outgoing sharemilker still had his stock on the property moving them from paddock to paddock chewing off every blade of grass he could find. We were horrified!

We also learned the occupants in the house were the sharemilker's workers and they hadn't been told they were moving out! The owner made a brief appearance but said he was preoccupied with the sale of his town property, then disappeared leaving everything to us. The furniture truck arrived and the huge lounge in the house was emptied by the occupants so our stuff could be stacked in. Our daughter unpacked a couple of our boxes so the outgoing people had something to put their belongings into. They eventually moved out on the 1st June, but we had to wait for the commercial cleaners to clean the house.

Three days after leaving the Waikato we were feeling very disheartened. We had no clean clothes, we couldn't get access to our furniture and there was hardly any grass on the property for our cows. There was a good supply of silage available so we decided we'd run all the cows in one mob and give them a paddock a day with four big bales of silage. I really felt for the young two year olds because they had to fight for their food amongst the older girls and lost a lot of weight. It was the only way we could build up grass cover on the farm before calving started in six weeks and we didn't have any equipment to do things differently, it was all in the shipping container. On June 4th our feed out hustler broke down so we were feeding the big bales out manually using the tractors front end loader and even that had to be welded back together a couple of weeks later! Was anything going to go right? We had so many *gremlins* in the works during the first few days!

We stayed in the holiday unit with our three cats for four nights. I was very conscious of our muddy wet gumboots on the doorstep of such an immaculate holiday facility and not being able to change into clean clothes. It was a huge relief to finally get into the house on June 3rd when we spent the day shifting each piece of furniture to its rightful place and finally got to sleep in our own beds. It was a lovely big home and we were spoilt for space. Our grandson had his own play room and we had enough space and beds to sleep seven extras for when family and friends came to visit.

The truck and trailer unit with the machinery and walk in chiller arrived without a problem but we weren't so lucky with the shipping container. It arrived later than scheduled on June 12th and we were given three days to empty it before it was collected. I was under a lot of pressure around that time and I felt I had to keep writing things down to clear my head.

Midnight by the Fire

Sleep is hard to find
The ideas are cluttering my head
Too many and I must get started
No! Wait till tomorrow.
The body is tired
Legs are aching, hands in pain
Enough for today
Wait for the sun tomorrow

The ideas are flashing still
The sofa is of little comfort
As the kitten wakes and begins to play,
Okay for him, he can sleep in the daylight hours

What is my hurry?
Why is my head racing ahead of time?
Surely I can switch off soon!
Keep writing and clear the head
This is good therapy
Write it all, then the brain can rest.

Tomorrow, fix the tractor
Finish the container and feed the girls.
Would love to finish the rock garden
Put in new soil and plant some seeds

Who knows what else will call me there…..
Too much to do and the days are so short
Make another list and make another plan
Those written down are those that are done.

Midnight. Eyes are burning tired
The body is beginning to relax
Let's embrace the feeling
And be grateful for everything accomplished.

What an incredible life I have
Every day full, no time for boredom.
I have to sleep now, my dreams await.
Okay kitty cat, off you get,
I'm going back to bed.

I had quite a few nights when I couldn't find sleep. I realised how burnt out I had become when I found the following: *So many Whys, too many Whys without answers. Dear*

God, how many challenges must I pass during one life time? I have always pushed the envelope and always had the strength to be outside 'the square', but I am tired now. Too tired to hope for a future. I have no wants, there's nothing to be wanted. I have my family, friends and my health, so just let me be. No more stress, no more worry, just be. Life's so simple, life should be simple shouldn't it?

CHAPTER NINE

Farming through Hell

In our first few days on the Northland property we cut down the huge creeper and privet trees that were overtaking the house which allowed a lot more light into the place. It was strange to see such mature privet trees, some with trunks as round as dinner plates in the house section. We had to mend several electric fences to keep the cows in and patch a number of leaky water troughs, some due to large cracks in the concrete and others with ball cocks held up with rocks so the water couldn't flow in. Even the water tank for the cowshed had a hole in it. It looked like a bullet hole and someone had unsuccessfully tried to plug it with some string. We began to blame the outgoing sharemilker for not keeping up with farm maintenance but maybe that wasn't the case. We had only been there a short time and I was hosing down one morning after milking our seven cows when the owner approached me and suggested if there was anything we needed for the farm, could we purchase it and he would reimburse us. I was a little puzzled. For some reason, against my better judgement, I reluctantly said 'That would be okay'.

A week later the landowner came to the house and said he was going overseas for three weeks and explained the sludge pump at the cowshed was unrepairable and would cost $3,000 to replace and 'Could we find an alternative solution'. 'Yes, we do have a pump that could do the job' but exactly what was the problem? We were getting less than half of the milk cheque and we were being asked to use our equipment.

Apart from those couple of times, we rarely saw the owner until he moved onto the farm in September bringing some milking goats with him. In our contract we had agreed to twenty goats but when we signed we didn't know the lay out of the property nor where the owner proposed to keep them, he just said they would be coming. When the goats arrived they were put into the two hectare paddock beside the hay barn which was used as their sleeping quarters. We watched as new kids were born and some of the older animals died. By October we had seen enough. The goats couldn't be milked because there were no facilities and they were costing the dairy operation by grazing valuable pasture. We volunteered to sell them on Trade Me which was acceptable to the owner and we collected the money from their sale and that of the mobile milking machine. Two days after the last goat was sold we handed over the money which we later came to regret.

For the first few months our main concern was looking after the girls, being able to feed them enough and producing as much milk as possible. Thank goodness we had decided to leave our weaner heifers with the grazier in the Waikato, because even though there was enough acreage, there certainly wasn't the grass growth. By the middle of July we had built up enough pasture to stop feeding out the baled silage. By the middle of August we had 75% of the cows in the milking herd and they were being milked twice a day. The last of the silage had been fed out by September 10th and we were still waiting for the extra grass growth. Ten days later I documented, 'Cows looking thin, grass cover not giving adequate food for them'. We were applying as much sea water and EM to the pasture as we could and we added AgriSea Animal tonic to their water troughs daily to ensure they were getting enough minerals. I had made four drums of fish fertilizer with biodynamic preparations but we couldn't use them for at least another three months. We looked to the owner for fertilizer but it wasn't to be. The property looked so *hungry* and we didn't have the money to improve the situation.

The mating season was only weeks away and I knew what would be ahead of us if we didn't get the cows fed properly. Our run of bad luck continued when our plan to run our two year old bull with the herd turned upside down when two weeks before his services were needed, he died of peritonitis. We began mating on October 10th using the AB (artificial breeding) straws we had in storage and then we let our eighteen month old bull run with the girls. He really struggled because he was so young. At that stage we weighed up the odds of either buying a

Hereford bull, which we couldn't afford, or finish mating altogether. We had only been mating the girls for four weeks but of the 150 cows, there were still thirty that hadn't been cycling! We decided to finish mating and cut our losses. If they hadn't cycled by then, they were not going to. We had a lot of empty cows at the end of the season because their overall health and stamina wasn't strong enough to carry a pregnancy or even produce ovaries. Many of them were only young. It was a heart breaking and expensive exercise caused by the lack of grass on our arrival to the property back in June.

When I said we couldn't afford a Hereford bull, we honestly couldn't. We had just paid off the last of our moving costs from the Waikato and we were paying all of the grazing invoices associated with the heifers we left with the grazier. I asked several times for our employer to pay his 60% share of the costs as our contract agreement stated.

Back in July we had to arrange and pay for a new water pump to be connected. We didn't have a choice, we had to have water. The pump serviced the house as well as the farm and I had been walking across the neighbour's property, approximately 700 metres, virtually every day to investigate the lack of water. I didn't need that! We got the experts in and found the resident pump wasn't strong enough to do the job and so traded it for a bigger one. The landowner reimbursed us a few months later. Another awkward situation involved the sale of raw organic milk. The owner bottled and sold it for three dollars a litre. We received twenty four cents a litre. The landowner explained that his raw milk business was up and running before we arrived and there was a lot of expenses to take into account. Our sharemilking agreement said all milk produced from the property was to be split 60/40 but it wasn't to work that way. My daughter and I felt robbed as the milk was taken from the vat every Thursday morning while we finished cleaning the dairy shed.

The monthly lease payment on our Whatawhata organic block was a big challenge that season but was also an advantage in the long run. We actually made a small profit by keeping the lease block that extra year, harvesting spring silage and then getting a second cut made into hay. Two truck and trailer loads were transported up to us in Northland and we sold the remainder to other organic farmers. It was the most beautiful supplement we had ever made. We were so pleased we had kept the lease block even though we had no stock on it. The drivers who trucked the bales north were amazed by its aroma and its lack of *smelly sludge* that they were accustomed to when they carried conventional supplements. They actually loaded cheese onto their trucks for the return trip and were very grateful for the clean silage bales they bought up. It was a relief to see them arrive because there was no way we could've made any supplements on the Northland property.

At the end of October we saw the clover in the pasture being devoured by the Clover Flea and it was eating every clover plant it could find. Many areas looked like a lawn mower had been over it. To me the appearance of the Clover Flea indicated the clovers were stressed and were sending out distress signals audible to the flea who saw it as a meal to be had. Plants low in nutrients and fertility are a beacon to a hungry insect and every plant of the natural world is ruled by this scenario. I see it as my role as a farmer to ensure the pasture is free from harm by feeding it with a nutritional fertilizer which would nullify its stress and deter pests. That place wanted food!

The first five months in Northland were extremely challenging but what was to follow was even more so.

Early in November our employer told us he was going to plant two and a half hectares in organic vegetables. He had mentioned the idea when we moved up suggesting he would pay us ten percent of the profits after costs. We declined the offer because we considered ourselves dairy farmers, not vegetable growers and personally I had my doubts the project was do able in land smothered in kikuyu grass. The vegetables were going to be grown and sold as certified organic so no chemicals were permitted to spray out the thick mat of kikuyu. The two and a half hectares were ploughed up ready for planting in November and the area laid barren until late April when it was dug for a second time and some vegetables planted. I was surprised by the plants chosen because I would never have planted them in the autumn, it was far too late. But this was Northland, maybe I was wrong, but time proved me right with the ground temperatures too cold and sunshine hours too short.

Looking back to November 20th I had written in my diary we were 'starting to see some grass ahead of us at last.' But then in brackets I had put 'Didn't last' and it was dated December 1st. That was a really quick turnaround. I remember being very worried about how we were going to get through the coming summer and the extra land taken out for vegetables didn't help the situation. On the last day in November we were speaking to the local bee keeper and he said there was no honey and the bees were unusually hungry for that time of the year. That made me feel a little better but didn't help solve the lack of grass. I made the decision to milk the cows every sixteen hours to increase our grass cover and consequently slow down the cows grazing rotation.

We began the new milking routine on December 4th bringing the cows to the shed at 5am and 8pm on the first day, followed by a twelve noon milking the next day. It wasn't a true sixteen hourly rotation but it worked really well and we managed right through till April 21st before changing to OAD (once a day) and then drying off May 1st. We had successfully used the same milking schedule on the Short Road property a few years earlier.

In October and November the landowner contributed towards his 60% share of the heifer grazing account and the supplements we transported up from Whatawhata, but the following month the payments stopped. Just after the New Year we had a meeting with the owner and we asked to update our agreement to the 50/50 partnership we had previously talked about. It was suggested we check with our lawyer about the legalities of our sharemilking agreement if the land changed ownership. We learned the Federated Farmers contract that we'd signed specifically covered our situation and we had every right to continue farming on the property if that was to happen.

On the second of February we sent an email to the landowner confirming our findings and suggested we should tidy up a few details in the contract such as compensation for the land taken for the vegetables and where the young stock would be grazed for the coming season.

We waited for a reply but heard nothing.

Amongst all the uncertainties there was one day when we were visited by two amazing *rays of sunshine*. We had the most unforgettable experience with Lynda and Jools Topp and their television series Topp Country. Through an article I had written for the Organic NZ magazine I was contacted and asked if we would like to film an episode about 'the love of organics' with Lynda and Jools. We felt honoured to share our passion and on February 1st we spent the whole day filming with a very talented group of artists. The day couldn't have been more perfect, the weather and the animals all playing their part just right, well nearly perfect. The camera man wanted a shot of the cows coming in for milking and they had to incorporate the sponsor's vehicle in the frame. The plan was for the Ute to be driven by Jools and Lynda behind the cows as they came to the shed. All was going to plan until the cows saw the strangers in the shed and before you could say *tiddly winks* they took off back down the lane towards the vehicle. 150 cows stampeding towards Jools and Lynda and we couldn't do a thing! Jools and Lynda jumped out of the Ute and began waving their arms and making noises and our daughter took off to cut the mob in half. Eventually we turned them and the camera shot shows me walking ahead of them. It wasn't our usual way of bringing the girls in for milking, but we got them there in the end. From a whole days filming we got seven minutes on television in episode eight of Topp Country's second season. We had to wait a whole eight months to see the result but it was all worth it. We were amazed with the coverage the producer achieved as the whole day seemed to be covered in that short seven minute time slot. If nothing else, our pride and love of the organic farming methods is now in the archives for generations to come, thanks to the Topp twins.

Topp Country was the ultimate distraction from what was going on behind the scenes. The following two months our employer seemed to be playing *cat and mouse* with us. I continued to forward the outstanding grazing account that was growing by the day but still received no communication about it or our renewed milking contract.

Early in March I caught up with him and insisted we all got together to sort out the contract as well as the outstanding account. The date for the meeting was set, as was the restaurant rendezvous. Every meeting we had with him was at a restaurant which was nice but different. The scheduled meeting wasn't to be a *meeting* but it was to be an informal and friendly talk over coffee between himself, his wife and myself. I wasn't comfortable with the idea of it being just me, hubby and our daughter should also hear what was to be said but they weren't included.

I had heard most of the issues before and as was suggested it was an informal meeting so by the end of it I felt I had gained nothing. The following day an email arrived agreeing with all we had spoken about saying he would sort out what he owed us in the following week. 'Hooray, sounds like I did make an impression'.

The next week came and went and so did the next, and the next. Hubby was getting anxious and so was I. The outstanding account was over $20,000 and with us only receiving 40% of the milk cheque we were barely living. That was the season Fonterra had to decrease payments to its suppliers and our cash flow was non-existent. We sent a text to the landowner, 'Pay up or we go the next step'. There was no reply. Then on March 31st, a couple of days later he came over to me and handed me an envelope saying 'This will get everything sorted' and walked away. I thought 'This is good. We have some money at last!' I walked into the house and opened the envelope. It was a two page letter but no cheque. My heart sank. The letter didn't really say anything significant, it was more on a personal note than a business one, but then the final paragraph said it all. We were 'dismissed because he didn't like our attitude'. Attitude? You mean to say, after waiting weeks for a reply to our email we weren't allowed to begin making demands and if we hadn't sent that text we wouldn't have been dismissed?

I was fuming! How can we be dismissed for asking for what we were owed? We had done nothing wrong. We had supplied grade free milk to the factory, cut down huge areas of old gorse, used our equipment to run the effluent system, tidied the house and its gardens, upgraded the dairy shed area, the calf shed and its surrounding fences and we had done all we could to help by accepting the 60/40 agreement and that letter was the thanks we deserved? I walked to the owner's house and the wife answered the door. I handed the correspondence to her and said 'This is an unfair dismissal and we don't accept it'. The landowner was down the farm so I preceded to confront him moments later. He had some feeble excuses which went in one ear and out the other and then I just let him have it! Personally I have never ever felt the urge to let off steam, but I had had enough!

From that day we slipped further and further into hell. We had never ever been treated like the scum we were portrayed to be by that employer. He explained to us we only had a tenth of the value of equity invested in the venture and so should only be entitled to a tenth of the income produced. That was the *attitude*. There was never an acknowledgement that without us and our animals the farm didn't have an income or a raw milk business. We were the scum of the earth.

Personally I started to search for an answer to *why* this was happening. I looked to my dreams, past and present for clues hoping I would find a direction. I had trouble remembering most of them but the very vivid ones were not easily forgotten. I had written down one dream: *A baby admired and cared for, a baby in a pram, but then the infant begins to shrink and turns into awkward shapes and clenches my thumb. It will not be shaken loose. The monster is now grasping my hand with all its strength.*

I remember that was the first of two very lucid baby dreams at the time. In my second dream, someone had taken a baby from a car and I was in tears because no one was looking for it. Both baby dreams were like no other dream I'd experienced. I actually felt the pain and woke with tears in my eyes. I did some research into what they might mean and found a baby can signify a new idea or a new project. I think I was unconsciously aware of losing my passion, (the baby) my organic status somewhere in the future, and it was going to be painful. At the time I didn't understand their significance, but I was about to.

Our Northland employer gave us six weeks to get off his property. To find another organic position in that time frame was 99.9% impossible. During the month of April we had two avenues we were investigating. The first was finding out if we could be dismissed like we were and the other was looking for new employment.

We were also very concerned about how were we going to get what we were owed. I made an appointment with a local lawyer and explained our position. It was difficult for him because we were new to the area and he didn't know us. He also knew nothing about farming but he looked through our contract and said there was a clause in it that allowed our situation to go to conciliation and then to Arbitration if not resolved. The information was emailed to our landowner who answered promptly saying 'No, the contract is finished. There is no contract and so the right of any conciliation doesn't exist.' The lawyer emailed several times and got the same reply. We were at a stalemate.

In the meantime we received an email from the landowner offering us the lease of the farm for $120,000 per annum. We estimated the property was worth $85,000 per annum to lease and even at that price it wouldn't be easy based on a five dollar pay-out for our milk. We emailed our expected cash flow to him but he insisted it had to be $120,000 or 'get out'. That was ten thousand dollars a month and we had to decline, it was too expensive. The next offer was similar but included 60% of the raw milk income to go to the owner of the property. It was still way above the true value of the lease and our figures showed we'd be making a loss of at least forty thousand dollars based on a five dollar milk payment. The Fonterra Dairy Company pay out for that season was three dollars and eighty five cents, a huge reduction from eight dollars the year before.

We were now into the month of May and the pressure to get off the property was enormous. We were instructed to shift all our farming equipment out of the implement shed so the owner could use it. At this stage everything said between ourselves and the owner had to have a paper trail so it was all via email. Every day I dreaded turning on the computer.

The grazing contract for the heifers in the Waikato ended on April 30th and the landowner took exception to us bringing the rising two year olds up to the farm on May 2nd. We were warned our stock would be sold and trucked off the land! We were stunned! Winter grazing rates took affect May 2nd and we had no alternative but to bring them home. We couldn't afford twenty two dollars a head per week and knew the chances of getting the landowner to pay his 60% share of the account wasn't good.

A small distraction for me at the time, was the annual Biodynamic conference. It was a breath of fresh air to get away from all the problems at home. During the first day I was sitting with a woman who had taught children at a Steiner School. In conversation I confessed to the pressure we were under and the unknown future we were facing. She left the room for a while and returned saying she had just sent us a blessing in the form of rainbows. I was close to tears. I seemed to be able to *feel* her energy and her ability to strengthen my will-power and I thanked her profusely as I left the conference. For the following three days I saw so many rainbows in the sky, I lost count. It was truly amazing! Tears streamed down my face as each one appeared in the sky. They gave me new hope and filled my heart with renewed strength. I was obviously supposed to attend the conference that day and experience biodynamic philosophy first hand. It was truly unbelievable.

Another magical event occurred when I phoned the Waikato grazier to organise our girls to be trucked to Northland. I explained our predicament and asked if he knew of anyone looking for farmers like us. He understood what we were going through and said 'There is an older guy out there for you. You have to attract him.' I was a little confused by the comment so I asked 'How are we supposed to attract this person?' He simply replied 'I can only see the headlines and there is an older guy out there for you.' I believed him whole heartily because a year earlier he told us the Northland position was not the one, and he was 100% correct, but now to search for an older guy 'Where do we look and how do we do this?'

CHAPTER TEN

At the Eleventh Hour

Gypsy Day was drawing closer and we had nowhere to go. The stress was near unbearable. During the final few days of May our call for support of any kind must have reached the universal spirit because we had so many *angels* call on us there was no other explanation! Nearly every day we had a complete stranger knock on our door for some reason or another who ended up having a cuppa with us and leaving us with their kind thoughts and best wishes. I couldn't describe their sudden appearance any other way, we had never had so many wonderful people, perfect strangers on our door step in such a short time frame, ever! Each visitor gave us new strength as we moved from one day to the next. We hadn't given up but the situation was draining us all enormously.

On May 29th my daughter and I returned from feeding the cows to hear that hubby had received a phone call from Federated Farmers and I had to ring Fonterra. I honestly didn't see the message as being important because I had contacted our dairy representative several times looking for any vacant employment positions. I wasn't in a rush and I recall it was about an hour later that I found the phone number and made the call. I finished the phone conversation and told hubby and my daughter 'Yes, there is a possible farm position in Hikurangi, just north of Whangarei and I have been given this number to ring.' I rang the number and the woman said they were waiting for another person to give them a yes or a no to the 50/50 sharemilking position but at that stage it was first in, first served. I asked if we could come for a drive and meet them both. Half an hour later we were in the car and travelling south. We had no expectations, we were just getting out of the place and going to meet some fellow farmers.

Arriving at the farm around one o'clock we introduced ourselves and went inside for a cup of tea. We felt at home straight away and Ethan was really taken by the little dog who was part of the household. Conversation came around to the sharemilking position we were leaving and they couldn't believe we had a 60/40 agreement where we got the 40%. They had never heard of such a contract. We had been in their company for only an hour and a half when the landowner announced 'Well if that's okay with you, it's okay with me and you can have the job.'

We drove away from the property feeling slightly stunned. What had just happened? We met this couple at one o'clock and had a new home to move to two hours later. How did that happen? We had experienced a lot of unexplainable moments during the previous week and this was another one. Dare I say someone was *watching over us* and am I to understand this was the *older guy* we had to attract?

May 30ᵗʰ was a Saturday and I had our lawyer's private phone number so I called him and told him the good news. As any communication with the Northland property owner had to have a paper trail I left it to our lawyer to inform him of our impending departure. Up until that point we were staying on the property and were waiting for the landowner to conciliate with us because we didn't accept the unfair and in our minds unlawful dismissal. If the same events had occurred in any other employment situation the employer would have been made accountable by the employment tribunal, so why didn't we have the same rights?

Our challenge now was to find any available stock transport trucks and furniture movers. There was only two days till Gypsy Day and other farmers would have booked in their herd shifts weeks in advance. Our lawyer informed the landowner we would be off his property by June 10ᵗʰ to which he expressed concerns over the movement of our cows clashing with his stock arriving on the property from the South Island. We were stunned.

We began packing the household items the very day we found our new home. Up until then we hadn't packed a thing. In fact on our arrival in Northland I was determined to not see another box and so unpacked every last one!

All our farming equipment, tractor, walk in chiller, stock and furniture were off the farm by four o'clock on June 5ᵗʰ. Somehow the extra adrenalin kept us moving and we left the property in a much better state than when we arrived twelve months earlier. Looking back, we would have been in dire straits had we'd continued on that property. As it was our girls arrived in Hikurangi with a condition score of 3.7 (should have been at least 4.5) and we had already fed them all the available supplement on the property before they were shifted in an attempt to keep them in good condition. That was the second time we'd shifted with very skinny animals which was beyond our control.

Moving off the Northland property also meant we had been forced out of the Organic Certification we had worked so hard to keep over the previous four years. That was the lowest I had ever felt. We had been robbed of nine years of our lives. Our organic status no longer existed. My pride and passion was shredded into thousands of pieces. We had tried to help a fellow organic producer but it didn't work that way. Up until that stage we had managed to transfer our certification from property to property, but those days were gone, my dream had come true.

There is a belief out there that every challenge you face, makes you stronger, but I was not so sure any more. Our spirit, our passion, and our livelihood had taken a severe hammering. My health was also suffering. I was going to the chiropractor on a twice weekly basis for a very stiff neck due to stress, and sleeping at night had again become a constant struggle. I had so much on my mind. My head was in a spin looking for what to do next and I was always searching for the answer as to *why* this was happening. What had we done so wrong to deserve this?

At the time our biggest challenge after finding our new home, was to get the Northland property owner to an Arbitration Hearing as he was still in complete denial about the whole affair. Our lawyer finally managed to have a three way phone call with the owner and the arbitrator when he pointed out it was our legal right to have our case presented before the court because there had been no conciliation. He pointed out that both parties had signed a sharemilking contract making provision for conciliation to be actioned upon a dispute and if no compromise was made the Arbitrator would make the ruling. There was a lot of debate about the latter but finally our day in court was set for 29th September.

I spent many hours searching and duplicating emails and preparing a multitude of statements for the day. At times it was really difficult to explain farm terminology to the lawyer who admitted he knew nothing about dairying. Sitting at the computer for hours as well as setting up house and managing the cows on a different farm, was a challenge. We hardly had a penny to our name and the lawyer had already warned us it would cost at least $20,000 to defend ourselves in Arbitration. We began to doubt ourselves. Did we really want that kind of stress? Were we really on the right side of things? Finally we concluded it wasn't the money we were after, we had been treated like *scum* and we had every right to clear our name.

As with every move I had to change our address and redirect our mail. Somehow the second rural delivery form I filled out didn't get processed and our mail didn't arrive. I knew our mail was being held at the post office for the first ten days because I had requested them to do so but by the beginning of July nothing had been sent through. I phoned them and they said it was on its way, everything had been stopped until they heard from us. Some mail started to turn up but there was still six weeks missing. I filed a written complaint and the post office looked high and low but found nothing. I contacted the rural delivery person who said she had delivered a bundle of envelopes and had continued to drop off mail at our old address because it had been collected from the mail box. She thought we had gone back to get it. Unfortunately we hadn't and we came to the conclusion the owner would have collected it for us. It was the middle of August and the mail was still unaccounted for, we were baffled. Surely it would have been returned to sender by then. The rural delivery driver made several attempts to contact the owner of the property without success. I finally asked our lawyer to send a letter asking for it to be dropped off at his office. Within two days a box of mail was dropped off at his reception, nearly three months late. The post office rang and I happily informed them the mail had been delivered at last. They said it should never have happened and it was an offence to interfere with the mail.

Meanwhile our new employers welcomed us with their friendship and understanding. In all honesty it is very difficult to describe the fabulous working relationship we developed in such a short time. It was a pleasure to be able to laugh together and work *with* the owners not *for* them. It was unbelievable to see the owner arrive at the shed every morning to hose the yard down after milking and help where he could, something we had never experienced before. It was a joy to get up in the mornings. All of a sudden we got the feeling we weren't alone and we were appreciated for our contribution to the farm. What an awesome place to be.

As I prepared for Arbitration the approaching spring was to be a bit of a challenge. Our new employer had originally decided to farm beef bulls having not filled the sharemilking position and then we showed up at the last minute turning everything upside down! So there was no conserved pasture for the dairy herd and only a small number of wrapped silage bales. We all knew the grass situation would make things difficult but the main ambition was to get through the season as best we could.

We milked the herd once a day until 10th October when the grass was beginning to grow so we moved to sixteen hourly milking, which is three milking's in two days and not the customary four. Our next challenge was getting the girls pregnant again. Most of the herd had lost weight so it was no surprise we ended the mating period with more empty cows than normal because as the girls carry their unborn calves, their eggs are developing for the following pregnancy. The lack of feed over spring didn't help and lead to the inevitable poor conception rate so we ended up milking forty cows over the winter. It wasn't their fault they didn't fall pregnant, it was ours.

29th September arrived and my daughter and I sat in the Arbitration court to hear our case being presented by our lawyer. The Arbitrator required all evidence be forwarded to him before that day so really there was nothing to do except answer any questions with a yes or no and sit and listen. Our lawyer did a wonderful job and by the days end we had won the case without a doubt. We finally got justice and proof that our dismissal was unjust and unreasonable. The arbitrator ruled we had to agree on the amount of compensation between ourselves because if he had to 'make the ruling a certain party may not be willing to accept it'

Our lawyer had the position of being the mediator because our ex-employer didn't have a representative. The full sum of our costs and losses was $40,000 but half an hour later that had been negotiated down to $30,000 to which I reluctantly agreed. It took us another hour to finalise everything and when it was all presented to me to sign I began to wonder if I had really won the case. Our agreement was presented to the arbitrator who advised both parties he was equivalent to the high court and what was signed in his presence would be classed as a high court ruling and was to be treated as such. When it was all finalised, the owner shook my hand thanking me for being fair, to which I abruptly replied, 'it wasn't about the money'.

Back on the farm, we were beginning to relax and allow ourselves to take in the beauty of the property we were privileged to be on. There were amazing bush areas with beautiful Kauri, Totara, and Puriri trees, and frogs! Wow, I hadn't seen frogs since my childhood days in the Wairarapa and there I was sitting on the hill, listening to them and watching them sunbathe on the sides on the pond below. It was awesome! On some occasions they were even jumping from under thistles and ragworts we were chipping. It was fantastic to see so many! In other regions our wonderful frogs have been pushed out of existence as the swamps have been drained and the land planted in grass. Back on River Road we had tadpoles in a fish pond but they couldn't change into their adult form and ended up dying. The world's frog population is declining in many countries. Scientists are blaming the sun and our rising temperatures but what about all the chemicals out there? Metamorphosis is a very delicate process and even a minute chemical interruption is harmful. It looks like I was destined to move to the northern part of New Zealand to bring frogs back into my life. I love them, they are so precious and so vulnerable.

Our first season was a challenge but it wasn't long before we felt we had found our home. We were living on a fantastic property and working with a magnificent family in a place we were meant to be. It just took the universe a while to *get all the pieces in the right place* for us to get there.

Gypsy Day, June 1st 2016 came and went and we didn't have to move anywhere. We didn't have to pack any boxes or load any cows on to trucks. We didn't have to change phone numbers, addresses or dairy supply numbers. We didn't have to change a thing and it was a fantastic relief.

As I've added on the years I really don't know how I came through the previous four seasons. I do know it really helped to write my emotions down when times were tough and our daughter who has played a big part in being able to continue my passion. She has an incredible love, respect and understanding of the land and animals, a rare quality to be proud of.

We milked more cows that second season and although we didn't have the best of starts milking forty animals during the winter, it was better than our first one. My daughter and I have milked cows every day for the past six years and we are looking forward to a holiday this winter.

Holidays and time away from the farm haven't been possible for the last eight years. Only once in our lives have we employed a relief milker so we could all attend a family reunion in the year 2001. Usually one of us would stay behind to look after the girls.

Our present employer is without a doubt, the older gentleman our Waikato grazier told us to attract. How our grazier knew and how it all came together has been a miracle. We now have a ten year contract and we're able to continue with what we know and do best, farming. With the owners encouragement we've continued to make our own fertilizers and EM (Efficient Micro-organisms) and spread them over the property. They have commented on how good the pastures are looking and give the credit to our fertilizer programme. It is a fourth generation property so they are in a good position to witness the change we have brought with us. We've planted a new fruit orchard and the ongoing joke is we have to stay around to eat the fruit off them, at least ten years!

Eighteen months after our shift to Hikurangi I experienced the most wonderful dream. It was another baby dream where I cradled it firmly in my arms and there was a lot of admiration from others around me. The baby wore a blue woollen hat and I was lovingly stroking the baby's cheek. No one was trying to kidnap or grab it this time and I felt calm and serene. I interpret my dream as suggesting my life is back on track and no one is going to take my soul's purpose away from me again.

By moving back to a conventionally farmed property we have been reminded of why it was so important for us to retain our organic status. We've now moved back to the beginning again when the land and pasture goes through a detox flushing out excess potassium and returns to naturally processing nitrogen from the atmosphere. We've been able to compare conventionally raised dairy cows alongside our organically farmed animals because the landowner asked us to milk some of his. We made some very interesting observations. I noticed our girls had beautiful shiny coats with white patches that were really white. The owner's animals were easy to pick out because their coats were still rough, their white patches were dull and some of them looked older than their age. We also compared their overall health and discovered it was easier to cure our girls than the others.

At each milking I convey my admiration to our oldest cow in the herd, number ten. She is fifteen years of age and still going strong. She has had mastitis only once in her life, and that was on the River Road property, has had a calf every year and has accepted her new surroundings as we've trucked her and the girls from property to property without batting an eyelid. We have milked her in seven different dairy sheds, on seven different properties. Some were completely flat, others were on rolling hills, and then there were the steeper hills on River Road, Pirongia and now Hikurangi. She's an incredible animal.

A recent addition to our household has been our grey Labrador. Our daughter said she wanted to name her Gypsy 'because we are not going to be gypsies any more'. We have shifted too many times and it is time to stop the repetition.

Who knows what the future holds, but as long as we have family, love and laughter we are rich.........very rich.

PART II

UNDERSTANDING THE GIFT

CHAPTER ELEVEN

The 2015/16 Season

We shifted to Hikurangi with a rip and a roar. We had to learn the new farms layout, give names and numbers to paddocks so we could identify where we were talking about and get used to working in another milking shed. The property owners weren't used to seeing organic management so secretly we had a point to prove.

One of my first priorities was to set up a drum of EM in the old vat stand and leave it to brew for between five and seven days after which I repeated the process in a second drum. We already had a hundred litres of fish fertiliser made up so we started to feed the pasture as soon as the weather permitted. The best time to apply fertiliser is when there's a bit of moisture around either first thing in the morning or during the afternoon around four o'clock.

Three days after each paddock was grazed the motorbike was used to spread a mixture of EM, fish fertiliser and chamomile. When the first round was completed, we began the second round of fertiliser using

EM and AgriSea. By the time the cows began to calve each paddock had been fed twice.

Occasionally we add copper, silica or selenium to the fertilizer mix. Copper in the autumn, selenium during the winter while silica can be added anytime during the year. All of them are in homeopathic form. The Hikurangi district was renowned for copper deficiencies so for four days we added a capful of cuprum, homeopathic copper to the herd's water troughs.

Approximately six weeks before the cows were due to calve we added Calmag, homeopathic calcium and magnesium to the herd's water supply to help prevent milk fever and we continued the practise until the end of calving. We turn to homeopathy many times during the spring period for mastitis, retained membranes, and sick calves. We also like to use the homeopathic alfalfa and minerals to boost milk supply and help the cows put on weight.

The young calves and their mothers are bought to the shed in the afternoons, the cows being added to the herd and the calves put into the shed. We don't take any milk from the mothers until the following morning. They are so busy looking for their babies they forget to eat and with empty stomachs the older cows are good candidates for milk fever. The following morning their quarters are checked for mastitis with the appropriately coloured tag attached to their tails. The freshly calved cows are colour coded with either red, green, yellow or blue insulation tape. That gives us four days and on the fourth day we return to the first colour, put their milk into the factory supply and use the colour for the freshly calved cows. The first four days of milk is collected from the cow and given to the calves and when her SCC (somatic cell count) is at the acceptable level we harvest her milk for the factory. Occasionally an animal comes to the shed with mastitis and we have to dry off the infected quarter because her somatic cell count doesn't decrease after we've cleared the infection. The following season the younger cows generally calve down sound in all four quarters without any further problems but the older girls can be more difficult to establish a long standing cure.

Before homeopathy we used to have penicillin to treat any infection, be it foot rot, mastitis or retained membranes, three days of antibiotic jabs was used for every ailment. Now we deal with many homeopathic solutions and the one we use depends on the signs and symptoms of the problem. Pyrogen, Bryonia, Gun Powder, Hepar Sulph, Belladonna, Phytolacca, Ipecca, Tea Tree, SSC, and MA Nosode are a few of the solutions in our medicinal cabinet that we call on when mastitis is a problem. It's a whole new world and one that we have had to learn because there is nothing else to turn to for organic dairy farmers.

The young calves are fed six litres of milk per day and under organic rules must be offered milk until they are at least three months old. We feed them only milk, grass and hay and if they don't look right for weaning at three months, they are given milk for longer. Occasionally a calf may show signs of lagging behind so we give her a spray of Calcphos (homeopathic) on the nose or vulva to strengthen her metabolism. We have witnessed many magical results over the eleven years we've used homeopathy.

Nearly all the pastures are starving in spring. They are chewed down faster than they can grow and it feels right to feed them as often as possible. We normally achieve two or three applications per paddock between July and October. Comfrey tea with added biodynamic preparations is another fertiliser we use to provide a food source for the foliage. I don't expect to *see* a difference in the pasture short time but long term I know the microbes are being fed and the work is being done underground.

In September that year we had two cows with peritonitis. They must have picked up an old staple or a piece of wire while they were grazing. The first case died within two days, there was nothing we could do. The second cow developed peritonitis about a week later. We put her into a paddock by herself and used every homeopathic solution we could think of. We used Aconite to get her temperature down and then used pyrogen and gun powder alternatively. By the seventh day we thought we'd get the vet out to check our diagnosis and he discovered we were right. He said he could give her some antibiotics but said she would probably die anyway. He couldn't understand why there was no temperature but he didn't want to know that we controlled it using homeopathy! Our girl survived her ordeal and lived to tell the tale until she fell down a cliff and broke her back. We just about cried, she had been through so much.

In early October, two to three weeks before mating

was due to start we added pyrogen (homeopathy) to the herd's water trough for four days to *clean out* the cows ready for the bull. Our three Friesian bulls were used one at a time, alternating them every one or two days depending on how many cows were cycling. By hand rearing our own bulls we've found them really easy to handle and draft out of the herd when they come to the shed at milking time. After a few days they learn the routine and we only have to call them and they'll come out of the yard.

There wasn't enough grass cover in early spring so we milked the herd once a day until October 10th and then we changed to every sixteen hours. Each season I have a book or a calendar where I record the date the paddocks are grazed making it easier to find the ones due to be fed off. Our grazing methods are the same as before we changed to organics but management of the herd's health is completely different. There are no treatments given without a disease being present and there is no *slash and burn* paradigm to *kill the bad bugs*. We look at the environment as a whole and discover how to change it to deter the unwanted scenario. Meantime in the homeopathic world *like treats like*. If an animal has swelling we use Apis which is bee venom. Pyrogen is used where there is a rotten smell or situation and it is made from rotting meat. Homeopathy is an amazing concept and there is no milk withholding period to worry about and it really works. Our cupboard at the shed contains at least forty homeopathic solutions, we love it and would be lost without it.

Each year we plant a summer crop with a mixture of turnips, chicory and red clover. Nature doesn't believe in monoculture and will naturally grow a variety of species together to complement each other. With our mix the red clover supplies the nitrogen, the chicory provides zinc plus numerous other minerals and turnips are the cow's source of protein. With only turnips in the crop, the stock lost weight because there was no carbohydrate whereas with our combination we didn't have that problem. When the crop is initially grazed the turnips get demolished but the chicory and red clover grow back. In the autumn the paddocks are under sown with some ryegrass, cocksfoot and white clover. There is no urea used because we feed the young plants with EM, AgriSea and fish fertiliser with the latter providing the nitrogen. Seawater is another fertiliser the young plants love. We fertilise and feed every new crop at least a couple of times in its early growth stage with the understanding a young seedling is like a baby and babies are always hungry.

At the end of November we started to see a few ticks on the girls so I increased the fish oil in the teat spray and every milking we knocked them off the cow's udders and stomachs and sprayed the BioSea where the ticks had been. Fish oil is an amazing healer as well as a tick deterrent. Every tick we squash prevents another two thousand being added to their population and I read ticks hold the world record for surviving without food for the longest period of time, a remarkable forty years!

Pulling ragworts and chipping thistles can be a challenge and we always try to get them done before the weather gets too hot. Our motto is ***don't let them seed.*** We always have a couple of chippers attached to the four wheel motorbike and as the cows leave the paddock for milking the ragworts and thistles are either dug out or chipped off at ground level. As I cut down a thistle I compare our method with the chemical sprays used in agriculture. Many tiny creatures live under and around ragworts and thistles and they would all be dead using the latter technique, the snails, slugs, honey bees, dung beetles, centipedes, spiders, slaters and frogs to name a few. These innocent creatures are living where there's a food supply and are part of an even bigger eco-system keeping our soils alive. Unfortunately I destroy part of their habitat but at least they are still alive to search for another one.

By Christmas time most of the young heifer calves are weaned and spread out two to a paddock. We have discovered the calves don't pick up intestinal worms as often when they're separated and there is less competition for food compared to grazing in a larger mob. If we do keep them together we plan their paddock rotation to be twenty one days or longer so there's more of a chance any intestinal worm cycle will be broken.

During the season I like to have salt blocks for the animals and we use the Himalayan rock salt. These pink rocks are full of minerals so the animals get more than just salt. The cow should have access to salt all year round especially when they are grazing Kikuyu pastures and in the hot summer months.

We usually finish mating and take the bull out of the

herd at Christmas but that season it was the middle of January after twelve weeks. The spring had been tough on the girls so we decided to leave the bull with them for an extra twenty one days.

After a dry November and December, the months of January and February were wet. On New Year's Day we got 100mls of rain and the next day another 35mls. Yippee! What a fantastic way to start 2016! The fourth round of fertiliser went on and we continued to milk using the sixteen hourly routine to conserve the grass just in case there was a dry spell ahead of us. That wasn't to be because on the first and second day of March the heavens gave us another 70mls of rain. We managed to make three paddocks of silage in late autumn which was a fantastic bonus.

We didn't see any symptoms of facial eczema that year. It used to be our old enemy but we don't fear it any more. The homeopathic FE nosode solution is put in the water troughs and if an animal does pick up the spores and show signs of facial eczema we use chelidonium to cleanse the cow's liver and support her recovery. We don't have to use any boluses (slow release capsules) and worry about them breaking open in the animal's stomach and we don't have to drench the herd daily for a couple of months with zinc that depletes their copper and who knows what else. We are very lucky to have such an incredibly easy and effective method to work with. The young calves are spread out and more difficult to treat and occasionally the odd one will get facial eczema. After four days of chelidonium however, they are on the road to a complete recovery.

By the end of April we knew we had more empty cows than normal so we decided to milk forty of them through the winter months and continue sending milk to the factory. We were also able to rear twenty autumn bull calves using the watery milk at the beginning and end of milking which would've gone down the drain otherwise.

By the beginning of May the younger cows were dried off leaving the older girls milking for another couple of weeks. Between the ages of two and four years, cows lose their baby teeth to be replaced with a second set. I used to think the process was over by the time the animal reached three years of age but I was proved wrong when a vet checked the mouth of a cow close to four and showed me a couple of gaps waiting for the next tooth to fill. With this in mind the younger cows get priority at the end of each milking season as we dry off the herd.

A large number of non-organic farmers insert an antibiotic into the cow's teat canal as she's dried off to guard against mastitis and some use a plug made from an antiseptic wax. We don't use anything. The odd cow does develop mastitis but we keep a close eye on them and add pyrogen to the water. We never return them to the shed to strip out the quarter like the professionals recommend. We've learned it encourages the udder to replace the fluid with more milk and the healing process doesn't happen. I used to strip dry cows for weeks without success but now we simply treat them via their water and within a few days the swelling recedes and we know the animal's body is healing itself.

To date it looks like we've made a positive impression with our organic methods, as we've heard complimentary comments from our employer, the neighbours and people familiar with the farm. Personally I am very proud but its early days yet. We saw the reward for our efforts on the Whatawhata block farmed under organic certification for seven years and know we have to be patient while the whole system detoxes and is released from its chemical history.

As I've caught up to the present day in my journey I realise I have used the first chapters in this book to vent my frustrations on how the agricultural world really works. I wish it had been a little easier but I have travelled through it recording just as many positive experiences as negative. Firstly I had the unique opportunity to compare my organic methods and discoveries on several properties and secondly I learned to have faith in the universe. I learned if we cared and supported each other more and took a stand against materialism we would be much happier.

To forgive and forget is life's biggest challenge. I readily except the past is in the past and I thank those who played a role in my journey for their participation. Without them I would not have found the pathway to where I am today. In paradise!

CHAPTER TWELVE

Unwrapping My Gift

My journey has had many twists and turns but in the process I have nurtured a wonderful partnership with Nature. It was important for me to begin at the beginning of my life to authenticate my passion, my suggestions and the farming methods I came to use. I also knew there had to be a level of understanding into how I was shaped thru out my life by my experiences on the farm and raising a family. I make no apology for the repetitions because somethings have to be read at least ten times before they're noticed. My deepest ambition is for my story to help mend the widening gap between farmers and the population living in our cities. A farmer's life is difficult at times and while mine has had its challenges, many others have experienced worse with earthquakes, floods, fires and long droughts. We need each other. Farming is the oldest profession on the planet and nutritional food is a necessity of life.

As I've been sharing my story I have tweaked the title of this book several times. I have been searching for the appropriate words but so far they haven't come to me. This morning as I woke at the usual five o'clock I heard someone tell me '*You have been given a gift'*. The message didn't come from a dream but it clearly came from somewhere. I kept repeating the words as I got out of bed so I wouldn't forget them. As I left the house to milk the cows I was surprised to hear myself singing. At first I didn't recognise the tune but once I got to the end of it I realised it was my grandfather's favourite song entitled Whispering Hope. I had not sung it for many years and I didn't know how I came to be singing it so early in the morning. I sang the song's first line again, '*soft as the voice of an angel, breathing a lesson unheard….*' Wow! Had I been sent a message from my grandfather and if so, maybe I should be taking it more seriously? The message had obviously been sent for a purpose and I had to find the reason why and what it meant. As I did my daily chores around the house and on the farm I searched for the gift I had been given. Nothing seemed to give me the right answers. I then went to google, wonderful thing that google! My investigation took me to Amazon where I've purchased several books and there I typed a search for *Gypsy*. A book called Gypsy Energy Secrets instantly popped up and I couldn't believe it but there in the introduction to the book was the answer I'd been looking for. *'Give yourself the gift of merging with nature'*. I was stunned and overwhelmed at my discovery. I then realised I have been given a gift and that gift has guided me to my amazing partnership with nature.

I originally used *Gypsy* in the title of my book referring to the June 1ˢᵗ shifting date for New Zealand dairy farmers but my search for *the gift* revealed a more meaningful reason. The further I looked into Gypsy Energy Secrets the more I noticed the similarities between my present day farming practices and the beliefs of those born within the Gypsy community. I read how Gypsies spend their entire lives in nature, walking on the earth, observing the seasons, and embracing life. Their philosophy is to make time to connect with the energy of the earth while every part in Nature is alive in spirit.

Gypsies call themselves the children of the sun; the sun is father, moon is mother, earth is grandmother and sky is grandfather. The teachings of the gypsy is to celebrate life and find the good in every challenge while leading a life full of passion. The book revealed I was right to look to the trees for guidance and embrace the power and strength of the mountain. Respecting their presence with a loving heart helps to heal our world and bring magic into our lives. A Gypsy knows when our living foods are replaced with the dead and artificial ones, illness will follow. We have a choice and can enjoy a full and happy life when we make the change and live in harmony with the earth.

How much closer to my story could a gypsy life be? I have followed my heart and I have been given a gift, the precious gift of pure love and understanding '*to be at one with myself and all my creation*'.

CHAPTER THIRTEEN

Who Stole My Money

In my opinion the worst thing ever invented by man was *money*. If we didn't have it, no one could steal it. Makes a whole lot of sense to me! The barter system of the past didn't make a monetary profit so something had to replace it. Maybe the money merry-go-round is the reason why the rich are just getting richer and the poor have become a lot poorer.

The worldwide organic movement has a voluntary group of people working alongside nature without receiving any money joining the endeavour as WWOOFers. They are 'Willing Workers on Organic Farms', people of all ages who offer their free labour in return for a cosy bed, fresh organic meals and learning the organic ways. It incorporates an amazing feeling that as organic we are all ONE, with the best of intentions to nurture ourselves and our earth and leave it in a better state than when we started.

Humans have come to owning and selling land for centuries, none of which really belonged to them in the first place. In my view the earth belongs to no one. We are only caretakers for future generations and are most probably considered a pest that has taken up residence on this planet called earth. Rudolph Steiner saw the surface of the globe as Mother Earth's stomach with the head being what is below ground. That presents the scenario that we are the inhabitants of the earth's stomach.

Looking back there were many corporate companies who in my opinion stole my money. Their intentions seemed honourable at the time but I think it was a case of us being hooked by their marketing ploys.

The use of urea on the pasture as a nitrogen fix was the biggest and worst additive to our environment which came at a price to the bank account as well as the soil. Over time we started to realise it was detrimental to both, but all the fertilizer reps and farm advisors championed the product and they were the experts so we never questioned them.

Artificial Breeding (AB) services really depleted any available funds and to this day I cannot see justification for the products we used. A large proportion of our herd has been naturally mated using the bulls we specifically chose taking the dame's statistics into consideration as well as the attributes of the sire. The bull calves were then chosen from those born within our herd with the occasional sire from the National Pedigree Bull sales purchased to bring in a new bloodline. I attended a seminar where a German scientist identified the interbreeding within the worldwide AB programmes. I wish I could recall all the details but I do remember he didn't paint the ideal picture and interbreeding could have been the explanation behind the two heifer calves we had to put down. They just never grew. Farming organically demands breeding bulls who are not propped up with drugs, chemicals or supplementary feed to sire the next generation. Over the years we've seen how artificial additives can weaken the herd and farm increasing costs exponentially.

In New Zealand there is one company that has the monopoly and holds all the records of a dairy cow's lifetime events, from the day the animal is born to the day they die. That too comes at a price and even for us with a smaller than average herd it equates to $2,000 a year every year just to have our girls on the data base.

Shifting costs have taken their toll over the years with the most expensive being our move to Northland which cost $23,000. Having to tackle four complete herd and furniture shifts four years in a row was devastating on our finances and they were during the years the payments for our milk was at an all-time low of $3.90 per milk solid. Maybe I should have given up farming then because it was a hell of a bumpy ride!

I have to mention the bank. Even though they have been very supportive over the years they have also been costly. I would've been a millionaire twice over if I didn't have to pay all the overdraft costs and interest on our term loans. The banks have made huge profits by financing farmers. They encourage them to take out loans to increase the size of their properties and increase production because the dairy company wanted an annual two percent increase in their milk collections. In a lot of cases farmers have had to subdivide and sell off small rural blocks to make any money. The whole system is *upside down*. The dairy company is a co-operative. Where has the co-operation gone since the merger of our smaller factories? The primary producer earns very little in comparison to those who collect their product and market it. What would the world eat without the farmer??

My Grandfather used to say that once in your life you need a lawyer, a doctor, a policeman and a preacher, but every day, three times a day you need a FARMER! —Brenda Schoepp

A professional that we all call on from time to time is the veterinarian. For animal welfare and our own peace of mind it is a wonderful service, but as the national herd has become less fertile and farmers have introduced so many supplements detrimental to the bovines gut, the vets are now sitting on top of a *gold mine*. Animal health costs have sky rocketed out of control. Many years ago when maize silage was first becoming the fashionable crop to feed dairy cows, it was a costly crop to plant and even more costly to animal health when lime flour (a source of calcium) wasn't added. Without it the cows suffer instant milk fever. We learned that lesson the hard way and lost two cows the very first time we fed them maize silage. The maize silage contributed to the runny, smelly cow pats and whole maize kernels left on the yard after each milking as well. I later learned only birds are capable of digesting the grains because they have a crop. The bovine's gut is designed to digest pasture such as hay, silage and green grass, not break down maize kernels.

Our vet bills were extraordinarily high back then and many times they were an expense that I now know could have been avoided by using homeopathy. When we made the change our animal health costs dropped from $26,000 to $2,500 in the first year.

By introducing the breeding bulls we personally knew the ancestry of, we virtually eliminated any calving difficulties the cow's had and our switch to organics took care of their infertility problems. These two management areas have cost us thousands of dollars over the years.

The annual leptospirosis vaccination was another event that didn't come cheap. After the vets vaccinate the whole herd plus all the young stock we'd receive an account for at least $2,500.The veterinarians are now asking farmers to increase the number of shots they give the cows because they have discovered the vaccine given in their first twelve months of life may not be adequate. It's a very expensive exercise, and one which we have to have or the vet can refuse to come to the farm for a sick animal.

Dry cow therapy involves inserting an antibiotic into each of the cow's quarters at the end of every milking season. It was an expensive and unnecessary exercise. The objective was to control and reduce the somatic cell count in the new season, but it didn't alter anything. We ceased the use of antibiotics in 2006 and depending on what property we farmed, there was no change to our basic somatic cell count in the milk supplied to the dairy factory. We compared our results with one of our neighbours who treated all their herd every year but their somatic cell count was never less than ours.

Another huge expense over the years was purchasing the heavy equipment required to work the high input farming schemes we found ourselves in. We had to have a bigger tractor to drive the forage harvester and tow the five ton feed wagon full of maize silage. We also had to have a tractor with a front end loader to lift the huge silage bales that replaced the silage pits and the bigger bales of supplement had to have a specialised feed out machine to spread it out for the cows. About the same time, the small two wheeled farm bike was being replaced by the four wheeled invention, more of an expense but a lot easier to carry gear on and jump off in an instant. They didn't fall over like the two wheeler either!

When we leased the dairy farms our expenses included the fertilizers. Before finding the biological and organic methods, we were spending between $10,000 and $20,000 a year purchasing solid fertilizers and using contractors to spread it on the land. What a waste! Looking at the properties we were on as well as the surrounding land, all we seemed to have done is encourage the yellow buttercup and California thistle to take over both the hill and flat country in an unstoppable wave. It never used to be there so why are they growing so prolifically and what are they really telling us?

Mother Nature will grow what the soil needs. If it is a calcium shortfall, she will grow dandelions, penny royal, and the lawn daisy whose foliage is calcium dominant. If the soil is compacted she will grow the dock, broadleaf plantain, and scotch thistles. The Californian thistle is prominent where there's *nitrogen pulsing* in the soil and can also suggest the calcium is not available to the pasture.

The simple properties of life are free.

CHAPTER FOURTEEN

My Body Can Talk

The further I moved away from the artificial world the more I noticed my increased sensitivity and the most unbelievable was my stronger sense of smell. I discovered I could smell the cigarette being smoked by the occupants of a car being driven down the road and I could smell the scented soap that had been recently used by a person walking some distance away. The cow's awesome aroma became more noticeable as well and when I gave one a hug it was the best medicine to lift my spirits. I also discovered the same beautiful aroma on the cat, the dog, and my grandson! It is a wonderful earthy smell similar to the forest floor and I see it as the sign of perfect health. During the winter months the cows have a different smell, it's sweet and milky but it's always a pleasure to inhale the air around them and be in awe of how beautiful it is.

A healthy body is precious to all creatures great and small. An unhealthy state can be noticed in a variety of ways. A mother for instance can smell her child's stale and unpleasant breath when illness is present. Sick animals also have a stale breath indicating something is wrong. As humans we tend to cover up our bad smells with toothpastes, mouth washes and anti-deodorants but is our body trying to tell us something?

In 2005 I had the experience when my body showed me signs of something foreign entering my skin and I saw a long line of black hair growing on the underside of my right forearm. I stopped drenching the cows but it took several months for the black hairs on my arm to thin out and change back to normal. Thankfully the black hairs have never returned. My real concern, after myself, was the probable impact the drench mixture was having on their milk. I couldn't see how it was entering the cow's stomach and not be added to their milk supply, after all a mother who is breast feeding has to be careful what she eats because it goes into her milk for baby, so why is the bovine any different? Further investigation into the drench I was giving the cows revealed it was a petroleum by-product. The manufacturers didn't supply a list of ingredients, only the benefits of using it. I couldn't believe it, a by-product of petrol to prevent bloat? Human food for consumption lists ingredients, why not for animals?

Since that experience I have kept a watchful eye for any distinctive black hairs growing where they shouldn't be. One such place was around the coloured area (areola) of my breasts where I noticed several clusters of alien black hairs. I didn't place them in the not normal category at that stage, I simply shaved them off and forgot they existed. I *cleared* the area a few times but it wasn't until five years later and our move to organics I realised the number had decreased dramatically and I no longer had to shave them off. In fact the reduction was amazing! Although it was a different scenario to the chemical drench entering my body through the skin on my arm, those black hairs were involved in the same process, eliminating foreign substances from within. A few strands of hair tested from our scalp can identify the toxins in our body because our hair follicles are the last area for foreign substances to accumulate before being expelled. The question for me was where had the toxins come from and how did they initially get into my body?

We are what we eat, drink and think.

Our move into organics has benefited us in so many ways but the elimination of chemicals, additives and preservatives has had the biggest influence. Since 2006 our groceries lists have included more and more certified organic products. In fact 95% of everything I either eat or use is chemical and GE free and I know this because the products I purchase carry the BioGro or Asure Quality logo. We are lucky to have our own organic cows to milk and eggs from our own chickens. They free range down the farm and are fed certified layer pellets. Here is a brief list of some of the certified organic food and household items we purchase:

Toothpaste	tea	pillows
Soap	coffee	duvet inners
Skincare	cornflakes	tee shirts
Dishwashing powder	rice	underwear
Laundry powder	herbal teas	
Toilet cleaner	sugar	
Deodorant	flour	
Coconut sugar	Himalayan salt	
Olive oil	Potatoes	

I know the introduction of the above products has made a huge contribution to me not seeing a doctor for the past eleven years. I last visited a GP surgery in 2008 for a badly twisted ankle. On that occasion the doctor on duty actually discussed my infrequent visits and we got onto the subject of today's nutrition, or lack of it and today's vaccination criteria. He commented that even his daughter, to his surprise, will not have her children vaccinated and she was a highly educated individual. The doctor I spoke to was non-committal either way on the subject of vaccination but did believe our modern junk food had a lot to answer for as he was seeing an increasing number of illnesses directly attributed to malnutrition.

It's important for us to grow our own vegetables so we can faithfully acknowledge what we are eating. There are too many unknowns in the fruit and veg aisle in our supermarkets. The Farmer's Markets are the best place but even then you have to question some of the growers because their produce is not always chemical free.

In 2010 I began making and drinking milk kefir. It

is a natural probiotic with a lot of the good bacteria we require for a healthy body. Good gut health promotes a stronger resistance to the colds, flus, and bugs doing the rounds. I also began fermenting herbal and flavoured teas with the kefir water granules. The water granules are different to the milk variety but both ferment their environment into an amazing beneficial beverage. They are a natural form of bacteria that feed on a couple of spoons of organic sugar during the fermentation process. The milk granules are even easier and only use milk as their food. They're both delicious and extremely good for you. The curd size of the kefir is smaller than yogurt making it easy to digest and especially good for people experiencing digestive problems. The longer you leave it fermenting the more B vitamins available but it does become a little bitter after the third day. Kefir can colonize our intestinal tract with several forms of friendly bacteria and can add beneficial yeasts that eliminate destructive pathogenic yeasts in the body.

Sour dough bread is another recent addition. It doesn't use any yeast grains to rise but instead introduces the environment's unique microorganisms into the dough. I use organic rye flour rather than the white variety to begin the fermentation process. Like the kefir it is an organism that has to be fed to be able to do its work and in this case the food is a little water and more rye flour. Sour dough is a heavier bread but our digestive system has no problem breaking it down.

Kombucha, originally from China is a popular drink in Russia and is a fermented black tea. The *mother* is a leathery mixture of bacteria and yeast and changes the tea to a sweet beverage with the help of a bit of sugar. The *mother* reproduces itself and before long there are plenty to share around. It is the same as kefir, the longer you leave it the more acidic it becomes. Like the kefir it sits on the kitchen bench covered with a mutton cloth so it can breathe and ferment.

The ginger beer bug is there as well and gets a spoonful of organic sugar every morning. All these natural fermentation processes have to be fed but the sugar used does not stay as sugar, the yeasts break it down into glucose and fructose and the bacteria convert it into simpler sugars, lactic and acetic acids, carbon dioxide and ethanol.

Since the introduction of living foods I have noticed how *silky* my skin has become, especially when I rub my upper arm against my torso, testimony to my new way of life. The skin is our largest body organ and works hard to keep us healthy.

Having successfully learned how to use homeopathy on our animals we have gained the confidence to use it on ourselves. Colchicum for a bloated stomach, mercurius for a sore throat, nux vomica for a sick stomach and motion sickness, aconite for a headache, these are just a few that I use. The philosophy behind homeopathy is to let the disease out of the body and not to suppress it. A year before we were introduced to it I suffered from several painful boils on my posterior. I blamed the dog for jumping on to the motorbike seat with dirty feet and me sitting in it with wet clothes. The boils were so painful

I had to go to the doctor for antibiotics. They all healed well but twelve months later another one as big as a bus formed on the top of my right shoulder. By this time my mind was opening up to the homeopathic concept and I learned a disease should be expelled from the body, up and out the top. I couldn't explain the unusual position of the boil on my shoulder because I had no porous skin there, no open wound or anything like that. It was a mystery except for the homeopathic concept of a disease suppressed has to leave the body at some stage, and maybe that was it. I didn't use antibiotics despite the size of it, only homeopathy while I nursed my shoulder until it healed. The boils have never returned.

Working with animals and being outdoors is my dream world, but it can be hard on the body. My injury list has included a fractured knee, broken little finger, fractured middle finger, tennis elbow, damaged right shoulder, twisted ankles, damaged wrist to thumb ligaments and carpel tunnel in both hands. Over time my injured body has self-healed and the only ongoing problem I've had to contend with has been the carpel tunnel.

My hands started to give me trouble fifteen years ago. My doctor told me 'A lot of farmers have the same problem especially when they are drenching the animals on a daily basis. It is the opening and closing of the hand as the drench gun is used to deliver the brew'. I was told I could have surgery but many I've spoken to say their problems come back so I wasn't keen on that idea. I have persevered with carpel tunnel for a long time and some days are better than others. When the cows are dry in the winter and I get a break from milking the carpel tunnel disappears. Occasionally the numbness in my fingers disturb me at night and it's difficult to find a comfortable position to keep the blood flowing through to my hands.

Just recently I stumbled on a possible remedy, magnesium. I am a huge believer in *we are what we eat, drink and think* and I am also aware that 98% of all disease is due to a lack of nutrition, or a specific mineral. Another symptom of low magnesium is the restless legs syndrome and having difficulty falling asleep at night, both of which I had trouble with. I began to add Dolomite to my breakfast yogurt along with some cinnamon, walnuts and roasted almonds. One teaspoon of dolomite is equivalent to 450g of magnesium, my recommended daily dose. The result has been amazing! My carpel tunnel has virtually disappeared and I am falling asleep so much easier. Dealing with fidgety, restless legs at the end of the day is also a thing of the past so my experiment has shown a positive result. I fill a jar from the bag of dolomite I use in a medicinal mix for the cows and it barely costs me twenty cents a month! I have since discovered coaches encourage their athletes to take dolomite tablets to prevent muscle cramps. For those unfamiliar with Dolomite, farmers also use it as fertilizer.

Sugar, flour and salt are known as the three white death ingredients of our modern society. For me personally my body tells me when I have exceeded the amount of sugar my body can take. Too much used to make my joints ache, especially in my hands and fingers signally it was time to quit the sweet stuff. Too much white flour would upset my bowel motions and contribute to a bad breath as it stuck to the insides of my teeth. The third ingredient on the list is salt. We should all be having salt in our diet but not those on the supermarket shelves. I often eat a raw carrot and add Himalayan rock salt. The natural salt has the same 84 minerals and elements found in our body whereas common table salt is composed of 97.5% sodium chloride and 2.5% chemicals and is difficult for our body to process. We offer the same Himalayan rock salt to the cows throughout the season in huge blocks for them to lick.

All three white ingredients never started out as detrimental to our health. They have been stripped of all their natural nutrition to look the part and over the years farmers growing sugarcane and wheat have had to add more chemicals to increase their yields and make a profit.

I am always trying to learn new messages by listening and observing my body and for four years I witnessed a hard lump on my nostril that regularly changed its size. I first thought it was influenced by the moon's cycle and then I thought it was a huge black head, but neither diagnosis was correct. When it decreased in size I happily expected it was on its way out, but the following month it would be back again. Thank goodness it wasn't visible to others but I knew it was there because I checked it most days. I decided to add mushrooms to my diet and I did some research into Dr Alla's MediMushrooms. I looked through the symptoms chart suggesting the health and medicinal benefits of mushrooms and was surprised with the huge range that was covered, blood pressure, arthritis, depression, cancers, impotence and asthma to name just a few. They were in tablet form and if I didn't want to swallow tablets I could open the capsule and spread the mushroom on my food. It was a cheaper option than buying them fresh and a lot easier than finding where to purchase them. I added three varieties, Maitake, Reishi and Shiitake, taking one tablet of each every morning. Three months later I got a huge surprise when I touched the lump on my nose, it suddenly broke open and released a clear fluid, the same you'd expect from a broken blister. After four years of waiting and wondering what the lump was, it became a watery blister and disappeared. It took another three weeks for the area to lose its tenderness and heal completely so obviously it went deeper than just the visible surface of the skin. I was amazed and can only guess the mushrooms had something to do with it.

Manuka honey is another edible capable of miracles. I developed a cold sore on my top lip and tried all the natural antiseptic creams and potions in the bathroom cabinet but nothing worked. I even tried colloidal silver without success. The solutions stopped the cold sore from growing bigger but it wouldn't heal. Into the fridge I went and put a dab of Manuka honey on it which had a MGO content of 100+. By the third day the tenderness had gone and by day five the cold sore had disappeared. The higher the MGO reading of the honey the better. MGO is methyglyoxal, a key compound in Manuka honey with the lowest measuring thirty and the highest 550+. When you're looking for the antiseptic cream in the first aid kit, this one works wonders, is safe and edible too!

I compare the health of my hair to the coat of an animal. The latter will be shiny, soft and silky when the animal is in good health and my hair is the same. I changed to using Eco Store's hair shampoo and conditioner several years ago. In the beginning their natural products wouldn't create a soapy lather because I still had many invisible and foreign substances being expelled via the hair follicles. As I decreased my intake of artificial and non-organic foods the shampoo began to lather 100% better. I have had my hair cut by several different hair dressers and been surprised by the number who made complementary remarks. My hair is all natural and I have never had anything done to it other than the regular shampoo and cut.

I am no medical physician nor chemist, I am a farmer, a mother and grandmother and by identifying the signs and learning to understand the ones I've seen, life has become a lot simply. There are signs everywhere and we all have the ability to decipher them. It is part of our DNA.

Our ancestors were a lot closer to nature and were able to make all their own decisions through observation. They knew what herbs or mushrooms were beneficial for a specific illness and what ones were poisonous. Such valuable lessons from the past are important and help us to stay in touch with ourselves and the natural world.

I share my story and experiences for all to enjoy, but my main ambition is to leave a worthwhile reason for doing so. Just as you watch the ripples spread over a stilled pool of water when a pebble interrupts its reflective surface, I see this book as being presented in the same way. If only one person is influenced by an aspect of my life, in any small way, then I have succeeded in nurturing and multiplying the essential needs of our universe. One person, one pond, one stone, and the ideas, like the rippling pond will grow exponentially to the millions.

The world has a bright future if we listen to the lessons of the past- Peter Jackson

PART III

THE LESSONS
I'VE LEARNED

CHAPTER FIFTEEN

Sharing Seventy Six Lessons

These are in no particular order and reveal why I chose to farm alongside Mother Nature and not against her.

1 I am absolutely certain pastures are on average growing less grass than they used to. In recent years many areas have struggled to grow grass in October early November and I don't remember having that problem thirty years ago. In my opinion this has been the underlying reason why farmers have tried to force the grass growth and apply urea as a booster. Thirty years ago we were harvesting silage in the middle of October without using artificial nitrogen. On each of the properties we've farmed September has always been the same, too much rain, not much grass and silage being fed to the cows for them to get enough to eat. I recall days in September when we put the girls into a paddock and by the time the last cow was milked and returned to the paddock all the grass had been walked into the muddy ground and there was nothing left. It was heart wrenching. A regular

September dilemma for us and many others is 'Where the hell I am going to put the cows tomorrow?'

2 It takes seven generations to change a DNA imprint. I have personally witnessed this after introducing organic methods. By the seventh season we had replaced the majority of the older animals with their offspring and we were seeing obvious signs that our animals had become stronger and were showing a greater ability to fight off illness. The more the animal's system is artificially supported, the more their DNA imprint changes. The same can be said with the introduction of additives and supplements to their diet. The more generations fed fodder other than grass, the more the DNA imprint of the animal's progeny will alter. It's evolution and not easily reversed.

3 A signed contract is supposed to be *sealed tight* but it is only a guide for both parties to work with. In our farming career we made every attempt to adhere to signed contracts, but disputes did occur and it was usually because a certain reference was missing or a phrase was worded incorrectly. I don't know how air tight you can make contracts without them being hundreds of pages long, and it seems that's the way everything's going. Bottom line, it is difficult working for another person on a property they own and if there is not some give and take from both sides, the agreed contract doesn't have a chance.

4 Be grateful for all that you have. There are many others with a lot less, no job, no home, and no family, nothing

to get out of bed for. There is always someone worse off than you. Several times I had to overcome my despair with those words and I had to focus on my amazing family and friends around me.

5 Don't follow the crowd. If everyone was to jump off the cliff, would you do the same? Not bloody likely! No two of us live in exactly the same place, farm exactly the same soil nor have exactly the same animals, so why does one rule fit all? It doesn't. We are all independent people and we live and work in our own individual spaces within our own capabilities.

6 Listen to your own *gut feeling* which is usually right. Back yourself and always think positive. Negative vibrations affect everyone around you and can drain energy from the need to move forward. Thinking and acting optimistically is the only way to enjoy life. Pessimism, looking at the worst in each situation will only breed more of the same and our auras will bear testament to that.

I find I'm always exhausted after going to the supermarket but it isn't because of the shopping it is due to people drawing energy from me as they walk passed. They unconsciously drag energy from my electromagnetic field or aura and I literally have to recharge my batteries after my trip to town. Animals are very aware of our auras and can instinctively tell when someone isn't authentic. Dogs are especially good at it. Our Labrador stopped a young guy at the gateway one day and wouldn't let him in. She had never ever done that before. I went to his rescue and he asked for some water for his car that had broken down. I obliged and he went on his way. I watched him return to his car and a couple of other people got in and drove off. That evening the real story emerged. The trio were staging the broken down car while they stole from the neighbour's house. The dog was correct.

7 Never, ever give up. I have never been a quitter and never will be. I can remember Dad telling me 'There is no such word as can't' and he was right. There is no such word in my vocabulary. That lesson has carried me through life and has been very valuable when times have been tough and not gone to plan. There's no such word as can't.

8 On every property our main concern has been to take care of the pasture to feed the animals. No matter where we've been, our incredible girls have worked hard for us. On a couple of properties they literally gave us the skin off their backs to put milk in the vat. Several times over the years I've looked at them and my gut feeling has told me 'our girls are doing this for us'. They obediently walk to the shed day after day and as long as there is food in the paddock they continue their task. I thank them after every milking. They are amazingly beautiful creatures who demand very little in return, maybe a hug and a scratch behind the ear. Thank you girls.

9 A bovine should be chewing their cud to a count of fifty or more indicating she has an adequate diet. The count will vary as the seasons change the pasture.

Her cow pat should fall to the ground in the shape of a perfect Pavlova with a reasonable consistency twelve months of the year, again proving she is well fed and has plenty of fibre in her diet. To achieve both of the above I learned cows should be eating at least a hundred different plant species during a five day period. Very few properties have more than twenty over their entire farm and that includes edible tree branches that a cow would reach. I have seen cows graze on certain grasses for medicinal purposes. On one occasion a cow close to having her calf raced to eat the dock leaves in her new paddock walking over all the rest. She was obviously going for the minerals provided by the dock that hadn't been available elsewhere.

10 Liquid fertilizer is just as effective as solid and we proved this when we leased the Willow Road property. We did soil tests every year and the following shows the test before we moved to the property compared with the results recorded three years later.

Analysis	31/03/09	16/03/12	Medium Range
pH	5.9	6.0	5.8-6.2
Olsen P (mg/L)	30	23	20-30
ASC (estimated) %	96	88	60-100
Potassium	0.65	0.61	0.40-0.60
Calcium	9.2	9.6	4.0-10.0
Magnesium	1.50	1.88	1.00-1.60
Sodium	0.17	0.14	0.20-0.50
CEC	21	24	12-25
Base Saturation %	56	52	50-85

Volume weight	0.81	0.65	0.60-1.00
Sulphate-S	23	18	10-12
Organic Sulphur	25	8	15-20
Boron	0.5	0.7	1.0-2.0
Available N kg/ha	184	230	150-250
Organic Matter %	11.9	15.2	7.0-17.0
Total Carbon %	6.9	8.8	
Total Nitrogen %	0.69	0.91	0.30-0.60
C/N Ratio	10.0	9.7	
AMN/TN Ratio %	2.2	2.6	3.0-5.0
Iron mg/kg	270	304	
Manganese mg/kg	484	522	50-400
Zinc mg/kg	13.0	15.8	2.0-10.0
Copper mg/kg	5.5	8.9	1.0-5.0
Cobalt mg/kg	1.1	1.2	2.0-4.0

A lot of the minerals tested showed their capacity increased over the three year term or stayed relatively similar. The volume weight decreased from 0.81 to 0.65 suggesting we had accomplished a more aerated soil structure by 2012.

11 'The best scientists in the world have four legs and a furry coat'. That statement was presented to me by Peter Bacchus, one of my peers who shared his experiences from a lifetime of farming and the results of the countless soil tests he had undertaken. The bottom line always came back to how the animals behaved, where they were happiest grazing and how they were managed. The soil tests only gave a small window into any situation that was still influenced by many other factors such as the weather,

nutrients in the pasture and even nutrients in the rain, Yep, rain is not just *rain*, it carries valuable minerals from erupting volcanoes, the cosmos and beyond.

12 Installed in our DNA is *a fight or flight* gene and it was pointed out to me that humans very rarely recall and remember the good times. It is the bad memories that come forward first. My body has been in the *fight or flight* space for many years as we were shifted from place-to-place and it wasn't until I joined a singing group in Whangarei that I became aware of how much my memory of the good times had been affected. After our weekly hour and a half singing session I would get in the car to travel back home but instead of recalling the songs we'd been rehearsing, my mind went blank the instant I left the hall. I couldn't remember a single word or a single melody! It was a very strange feeling. I could remember the music I sang with my aunties and uncles years before, but nothing I had just sung. It was twelve months later before my memory began to recall the music we had been practicing and at last I was singing the songs as I drove home. I was no longer in the *fight or flight* mode, I had found how to live again and remember the good times.

I wonder if cows have similar memories. I know our girls have a memory for things like the white bucket that had their milk in it when they were calves and the motorbike that used to deliver it to them. They also like to come into the milking shed in a similar position every day with the same leaders always being first. It has become a lot easier to draft individual cows from the herd while they're still in the paddock without having to bring them back to the yard. They seem to understand and remember what our actions mean. I believe as we are moving away from hurtful events such as injections and burning their horns off at a young age their *fight or flight* memory is mellowing.

13 Silica and carbon were the first two compounds to initiate the development of our planet and now the latter is being blamed for our climate change. There are actually two different types. The carbon emitted from our fossil fuels and vehicles is completely different to the carbon in the cow pat. Industrialisation can be blamed for our warming climate but not the carbon being emitting from the tundra's, the jungles and our animals.

Lead, mercury and silica are three of the many minerals delivered to the earth from the cosmos every day. Without a doubt we are influenced by the vast amount of star dust and elements in our atmosphere which are invisible to the human eye.

14 Homeopathy is a system of healing from the past, for our future. It was practiced by Hippocrates (c. 460-377 BCE) but Samuel Hahnemann introduced homeopathy to the modern world in the eighteenth century. It is safe and effective to use on humans and animals and allows the body to stimulate its own healing powers. It is so easy to use, isn't toxic to the environment and for the farmer it's fantastic because there is no meat or milk withholding time to worry about.

15 On nearly all the properties we've lived, I've had a list of paddocks the cows didn't like. No matter how delicious and sweet the grass looked to me, the girls would take one bite and walk out again. They were telling me something was wrong. Obviously the grass wasn't as good to eat as I thought. This common occurrence was very annoying. We tried shutting the gate and ignoring their moans and bellows but they would win in the end and get shifted to another paddock. They would rather stand at the gate and go hungry. Since applying our own fertilizing programme after changing to organics we have had very few upsets like this, giving me the confidence we are on the right track.

16 Supreme gut health is a must have for all mammals which includes us. It is the *engine room* that fights bacterial invasion and sets up immunity against everyday environmental invaders. I have turned to the humble mushroom and cultured raw cow's milk to feed my digestive system natural probiotics. I have added Shiitake, Maitake and Reisha mushroom capsules daily and each breakfast I add kefir (fermented milk) to my plate of fruit, yogurt, walnuts and roasted almonds. (Almonds must be cooked so our digestive system can break them down) The exact same *engine room* concept applies to our soil, the *gut* of Mother Earth.

17 I will never use antibiotics to treat an animal again, especially if a cow has Black Mastitis. As soon as an animal shows any symptoms we immediately treat her

with the homeopathic Nux Vomica and Pyrogen followed with gunpowder (homeopathy) and monitor her progress on an hourly basis. We do not strip out the infected milk as vets prescribe because her body will try to replace the lost fluid with more when she really has to stop producing and use all her strength to heal (milk is her medicine). Black mastitis is a very cruel infection and the animal never returns to the herd. Usually her udder will break open and weeks later the exposed tissue will dry up and fall away. I have seen seven cases in my lifetime. The first four cases were treated by the vet with anti-biotics and all died within days. They were so sore, they didn't want to go for water or walk the paddock for grass and they just stood there with their heads down. The last three animals got Black Mastitis after we had switched to homeopathy. All of these animals survived and showed very little pain as they dealt with the cruel disease. They continued to eat and drink through the whole process and never lost their will to live.

18 Urea and superphosphate promote a feeding ground for pests because they contain reducing sugars. A reducing sugar is any sugar that is capable of acting as a reducing agent because it has a free aldehyde or ketone functional group. The higher the reducing sugar content in the plant, the more carcinogens it produces. Superphosphate ties up the calcium in the soil and nothing happens without calcium. The plant will harvest Ca from the atmosphere when it has a brix (refractometer) reading of thirteen or higher.

19 The butcher who processed our home kills commented how he could distinguish between a slaughtered animal who had been grazing natural grass and those who had been eating pasture boosted with urea (synthetic nitrogen). He was able to leave our freezer beasts hanging for the recommended ten days before being cut up, but any carcases that came from a farm using urea would only last five days after which the meat would start to go rancid. He also mentioned the meat would stick to his knife making it more difficult to process.

20 There is a huge difference in the way our cows eat certified organic haylage compared to a conventionally farmed and harvested bale. As we have stepped temporarily out of certified organics we have offered non-organic supplement to the girls. They would rather starve than completely eat all the haylage we fed them. We haven't seen this since 2007, the first season we were in conversion to organics.

'The best scientists in the world have four legs and a furry coat'.
Peter Bacchus

21 The herbs and grasses in the pasture tell me what the soil is like beneath them. The daisy and buttercup families relate to calcium availability and the plantains relate to silica. The multi-mineral plants are the thistles, docks, gorse and broom with the latter two bringing nitrogen into the soil and like clover, they show this in little pink nodules on their roots. Self-heal, speedwell and chickweed are full of minerals and tend to grow where

the earth has been damaged. A note here that these three and the leaves of the plantain go nicely in the home salad alongside some garden fresh tomato and sliced cucumber! It's a good idea to taste their foliage first. If they are not sweet to the palate then they are not getting the correct food balance from the soil. Chicory is another herb we can eat but again, taste it first. If it's bitter or sour it means the ground in which it grows is probably the same and if it tastes awful to us it will taste the same to the cow.

22 Driving back to the Waikato from Tauranga on a summer night in 2003, we were overcome by the huge amount of flying insects that collided with the windscreen of the car. There was a constant bombardment as we drove through the agricultural areas but less as we drove closer to the towns. The insects out there are having a party at the expense of our unhealthy agricultural pastures which are providing the ultimate breeding and feeding grounds for them. Sick pastures are inviting insects to feed on them because the weak plant will emit a signal saying 'We are not feeling too good' and as Mother Nature commands, 'The weak shall be weeded out so the strong can survive'. Clover is a good example. The farmer has made it weaker and unable to produce nitrogen nodules on its roots. Clover is not required to perform its job in the soil where synthetic nitrogen has been applied to the land. Nature can supply plenty of nitrogen without our help, there's literally tons of it in the atmosphere above us.

A good way to discover who is in control is to answer the following questions. Who sowed all the docks in the horse paddock, where did they come from and who planted the willow weed on the old drain cleanings? What about the chickweed that takes over the young grass pastures and even the thistles; why are they there, and who told them to grow? Weeds like thistles and chickweed are succession plants and are the first to appear where the earth has been disturbed. Their role is to *mend* and mineralise the area ready for the second wave of grasses and herbs.

23 In 2001 I lost respect for the livestock breeding institutions in New Zealand when their representatives paid us a call. With our herd records in front of them we were told to cull the fifteen cows on the low production worth page. Her reasoning was 'They are your lowest breeding worth animals and are dragging the herd's average down'. I was dumbfounded! There will always be animals in the lower range no matter what the herd's average is and did they think we were made of money to simply get rid of fifteen cows? At that stage the herd was in the top five percent of the national herd average for BW (breeding worth) and PW (production worth) but since then I have questioned why the progeny from the bulls offered to us to breed from have not reached the standard promised and we have slipped down the ratings. Now it doesn't bother me. We breed our own bulls from our top producers, have next to zero calving difficulties and our success is evident in their strong constitutions, quiet temperaments and milk production.

I may be a little harsh in my impression of our

livestock breeding services because of my experience but I also want to acknowledge they provide a very valuable service to New Zealand's dairy farmers. My point is, as a farmer our livelihood is at stake and maybe we have put too much confidence in those behind office doors.

24 Despite scientific progress, our food is less nutritious now than before 1940. This is a scientific fact and as a nation of farmers originally set up to feed the increasing population of the United Kingdom, New Zealand should be worried. We have applied years of fertilizer but mineral levels in our soils have decreased – Calcium by 76%, Phosphorous by 34% and Magnesium by 50%. If the minerals are not in our soil they will not be in our grass, meat, milk or horticultural products. 98% of all animal and human diseases are the result of mineral deficiency. Science has clearly missed the boat and the statistics are in.

Senate document No: 264 records the declining mineral values in agricultural land around the world over the past century: North America 85%, South America 75%, Africa 74%, Europe 72%, Asia 76% and Australia 55%.

25 Kefir is a natural probiotic and I make the milk kefir as well as the water kefir for us all to drink. I started to make it at Willow Road and had the kefir granules multiplying so prolifically many of them were fed to the chickens. We shifted from there to the River Road property and for some unknown reason none of the kefir granules

multiplied for the whole twelve months we were there. They still fermented the milk and cultured wonderfully flavoured drinks but that was all they did. After moving from there, the multiplication started again and hasn't stopped since! I'm not sure what they were telling me but obviously something wasn't right, be it the milk, the water or the land, it was very strange.

26 Don't let them seed. A plant is a weed when it grows somewhere we don't want it to be. Each plant is there to play a role but if more are not wanted then chip them off before they can multiply. Some weeds may have to be chipped for a couple of growing seasons but by repeating the process the plant will become weaker and will eventually disappear. Most of our unwanted varieties are the succession weeds and after they have done their *job* they disappear. As an example, one of our turnip/chicory crops grew with thousands of thorn apple plants it in. They grow with a white flower that quickly becomes a prickly seed pod. We were devastated! We were organic and weren't going to use any spray to kill them so we started to pull them out one by one. Just when we thought we had them all, the next generation grew. We could just watch, there were too many of them. The new grass was sown that autumn and we waited to see how many thorn apple plants we would have to deal with the following spring. There were none! Not one in the whole paddock. We know we didn't pull them all out so what happened, why didn't they germinate? A similar incident occurred with a huge area of flowering ragwort.

My daughter and I spent two days pulling an area where thousands of ragworts had taken control. We pulled them out one by one and just left them on the ground where they had come from. We expected to repeat the exercise the following season but were absolutely amazed when we only found a handful. Again, what happened, where did they go? I know the ground must still be full of seed so why didn't they germinate?

27 Pesky biting flies haven't bothered our herd since we began using and incorporating organic and natural methods. Maybe it's because they only get grass to eat and their perspiration smells as a bovine's should. I realise humans have a foul sweat and disgusting bowel movements after eating dead artificial food so why should an animal be any different. Fill them with junk food and their foul odour will attract the flies. During the summer months I have compared our girls with conventional milking herds and discovered ours happily graze scattered around the paddock while the others huddle together frantically flicking their tails to chase away the flies. I've also noticed ours don't seek the shade of the trees as much either. They don't seem to be bothered by the heat of the midday sun.

28 Urea does boost grass growth but the plants will have hollow stems, it has been forced to grow. There is no substance, no fibre in the grass and the animal has to eat more to accomplish a satisfied stomach. I have witnessed our girls getting fuller bellies on less grass

since finding our new passion and their cow pats are the perfect Pavlova shape. The sugar level in a pasture can be measured with a brix meter (refractometer). When the grass shows a brix reading of twelve or more I know the cows are eating nutritional food. Fruit and vegetables have brix readings according to the variety measured and like the grass, the higher it is the better. A higher brix means less pests and diseases are attracted to the pasture. The brix is a guide to things moving in the right direction and can be influenced a lot by the weather and the time of day it is recorded. Milk can also be tested with a brix meter and on most days our girls produce milk with a brix reading of twelve to fifteen.

29 High cadmium levels in the soil are due to the application of phosphate fertilisers. This is of concern in areas around Hamilton where due to the high amount of cadmium some land cannot be used for housing and isn't suitable for growing food crops. Other areas in New Zealand such as Northland and Taranaki have areas where cadmium is also measuring above the allowable level. The worst scenario in the future could see farmers made accountable for the excess and be prosecuted. This is outrageous as all they were doing at the time was following the advice given to them by fertilizer companies and their field reps.

30 We must stop propping up the system. With the overuse of antibiotics, chemical worm treatments and vaccines we are creating an environment that cannot

stand alone. The industrial revolution hasn't worked for our land nor our animals, all we have done is chased the money rainbow and like that rainbow it continually moves as more interventional inputs have been included. It may already be too late to turn back the clock. Figures from a Dairy USDA survey suggest cow deaths are increasing year after year. In 2007 the largest cause of death was lameness or injury (20%) followed by mastitis (16.5%) calving problems (15.2%) and unknown causes (15%). I'm not sure if any statistics have been gathered in New Zealand but if the current drive towards factory farming is successful we will have the same problems as in the US. A few years ago I was speaking to a farmer who was very interested in organic methods and he described how he had recently been a relief milker on a large property where a lot of non-grass supplement was fed to the cows. When the owner returned he had to tell him there were three dead cows, they'd just dropped dead on the feed pad and he was mortified. He told the owner expecting him to be outraged but he replied, 'Not a problem. It happens all the time'. You might think the owners comment made the relief milker feel better, but it didn't. He couldn't understand how blaze the owner was. Those animals died for a reason and their owner made no effort to find out why, he didn't even care.

31 The fertilizer we like to use most is a mixture of BdMax Etherics @ 250mls/ha, EM (Efficient Micro-organisms) twenty litres/ha and fish emulsion. The first two can be found using the internet and the latter is a

simple recipe of fish skeletons immersed in rainwater with biodynamic preps for at least six months and then adding the amber liquid at five litres per hectare. The girls love it and will stop to lick the residual off the motorbike as they walk passed. Honey bees are also attracted to the fish fertilizer on the bike so who needs science when nature is recommending it?

32 Copper deficient animals show a brown hue in their coat, especially on and around the brisket area. Cuprum is the homeopathic solution we use when copper is required and we don't have to worry about giving injections. A small capful goes into their water supply each day for four days or for young calves we add it to their milk. Animals suffering with a worm burden signals a lack of copper in their body. Intestinal worms do not like living in a copper rich environment and ticks and lice are not readily attracted to an animal with a generous supply of iodine.

33 Cobalt (B12) is the mineral we use to boost young animals whose growth seems to become stagnated. Again we turn to homeopathy using cobaltrum and put a capful in the water or their milk. Selenium is another vital mineral we can administer using the same method.

34 Fish oil such as BioSea teat conditioner is always part of our medicinal cabinet to repel ticks and lice as well as repair cracked teats. On the River Road property we were grabbing handfuls of ticks from the cow's udders during milking and heaping them out on the concrete

for the birds to eat. Sometimes there were so many the ticks would start crawling away so I poured some fish oil around them and to my surprise they stopped in their tracks, they really didn't like it. Makes me think if we add omegas to the fertilizer programme could we lessen the tick population or keep them under control at least. We have noticed the tick population is less from the paddocks we have sprayed our fish fertilizer on but its early days yet. The girls really have to live with pests but to alleviate some of the infestations we rub the fish oil into their ears and around their udders especially where they can't reach for themselves.

Diatomaceous Earth (DE) is a natural silica mineral and a natural insect deterrent we use in the vegetable garden. It can also be used for animals, birds, orchards and crops and is safe to use anywhere in the environment. DE has 101 uses and originates from ancient fossilized algae provided by Mother Nature.

35 The media manufactures consent. If the media says okay then everyone will follow like sheep. This is a horrifying fact of our modern world but thankfully easy access to the internet gives everyone the opportunity to search out the truth for themselves. Endorsed by our media we have developed such a strong belief that saturated fats and cholesterol are bad for us and we have forgotten that in nature those are the exact same foods where the most nutrients come from. Hundreds of studies have proven how low cholesterol leads to much higher death rates from cancer, respiratory and digestive

diseases, especially for men with levels below 4.1mmol/L. As for women, the higher the cholesterol the longer they live. The cholesterol dangers are largely a myth and the deadly fats we eat are the manufactured ones such as margarine, homogenised milk, soy oil, canola oil, corn oil etc. The good fats are those grown by nature herself, coconut oil, butter and olive oil.

36 Antibiotic treatments don't stay in the animal's tissue, they are eventually excreted out on to the soil sterilizing the ground and destroying the soil microbiology. Chemical pour-ons and injections used to kill intestinal worms are also eco-toxic. It is written on the merchandise but few people read the fine print and not many care, they just want the product to do its job. Both of the above remedial treatments are coming to an end because nature is ultimately building an immunity to them. There are no new antibiotics to manufacture and scientists have had to alter the recipe of the chemical pour-ons and injections on many occasions to combat the growing population of drug resistant worms. We've found it extremely easy to stop using drugs and chemicals and replace them with non-invasive solutions such as homeopathy that don't interfere or destroy the environment they are used in.

37 Iodine is a very important mineral for animals as well as humans. The Feijoa fruit tree is a valuable source of iodine as is seaweed, the latter playing a huge role in our fertiliser. In the past our cleaning routine in the dairy shed included putting an iodine rinse through the

machines. Iodine is linked to thyroid function and a lack of it can lead to goitre. The cleaning products I refer to were banned a few years ago for no apparent reason. I have recently learned pharmaceutical companies lobbied for a reduction in the recommended upper safety levels for all supplements and in 2010 in USA were trying to set the upper safety level of iodine at 0.5mg.That is 96% below the minimum daily recommended intake! These big companies are manipulating the system so they get to sell their synthetic thyroid hormones, oestrogen suppressing drugs and chemotherapy. As new research confirms iodine is essential, another paper will come out saying it is dangerous, no one knows who to believe. Iodine is a natural substance and the cells in our body must have a plentiful supply to combat iodine suppressors such as fluoride, chloride, chlorine, bromine, nitrates and nitrites. We also require iodine to counteract the huge amount of electromagnetic interference coming from our computers and mobile phones. Iodine is a natural antidote for radiation. The only advice I recommend is to keep an open mind and discover the truth for yourselves.

38 Yellow Bristle Grass is one the latest invaders of our grazing land. The wiry thin blades of the mature plant are unpalatable for livestock and it has spread into pastures all around the North Island. We saw its invasiveness on the Willow Road and Pirongia farms. There was nothing we could do. Yellow Bristle grass is another noxious weed added to a long list which includes gorse, broom, California thistles, Wooley Nightshade, blackberry, and

many more. Science is out there to *kill* the invaders but hang on, shouldn't they be investigating *why* these invaders are growing in the places they do? Why has nature put them there? What type of environment do they thrive in? Why do they grow in some areas and not others? Is it because we have damaged the ground so much that we have destroyed a certain natural fungi or bacteria that doesn't allow a certain plant to grow or is there a native species that can provide an inoculant for the soil? Have the chemical fertilizers changed the soil allowing these foreigners in? I have many unanswered questions but I know to kill is not the way forward. Nature has put these invaders in place to do their *job* and we should investigate just what that job entails.

39 While living on Willow Road we experienced a very unusual event when the house was literally taken over by honey bees. We had to keep all the windows and doors shut for four days before the bees gradually flew away. They were not swarming they were just buzzing around the house with many settling on the bricks. We wondered if they were looking for food so we put out some saucers of honey and they instantly went for it. We did the same on the second day. A few of them looked sick and died on the decking. We decided to have a look at the farm's resident beehives that were 500 meters away from the house, and discovered thousands of dead bees at the entrance of the hives. The bee keepers had been there a few days prior to our bee invasion and obviously did something wrong. We rang them but never got a reply so

to this day we don't know the full story. I felt so sad for those little guys because it looked like human error had caused their unscheduled deaths. For them to choose our house as possible shelter was a real mystery because they had to fly over other buildings to get there.

40 We haven't docked cow tails for many years and very rarely have a tail flicked in our face during milking. When the rules changed and we were encouraged not to continue the tail docking practice there were many who insisted it was necessary to keep the cow's udder and teats clean. That wasn't the case for us. We have no trouble with dung on their udders and tails and can attribute this to them being all grass fed and not eating foods foreign to their metabolism such as palm kernel, maize, meal etc. Our girls produce the perfect cowpat throughout the season while a huge majority of NZ's national herd suffer from diarrhoea most of their life. You only have to talk to an AB (artificial breeding) technician and they will tell you how loose and watery the cows bowels are and how gritty the ones are who get fed palm kernel. There is a new disease out there and it's called *metal disease*. The symptoms arise when an animal has swallowed metal fragments while eating palm kernel and after the metal has entered her stomach, death is imminent, there is no cure.

41 Cow pats left in the paddock should break down and disappear within ten to fourteen days. In drought conditions this will take longer but generally if the dung

hasn't decomposed within that time the soil is lacking the correct microbes. Occasionally you may see a single cow pat still on the ground weeks later possibly indicating the animal who left it was sick at the time. The dog will tell you if the animals are healthy because they really enjoy eating a fresh, warm heap of dung! Another creature to show you things are progressing in the right direction, are the dung beetles. A big black variety was introduced to New Zealand about ten years ago and we have native dung beetles with a dark bronze shell. Recently I was told Northland farmers knew of a black dung beetle working in their soils several decades ago. All varieties only stick around if their food supply is *yummy*. I have heard a number of the introduced beetles didn't like the original environment they were released in so they moved over to the neighbour's property which happened to be organic.

42 Five basic principles of life: Never underestimate Mother Nature. She is Queen.

Always leave the door open for others to follow.

Always be willing to learn the lesson.

Everything and everyone are here for a reason.

For every action there is a consequence.

43 The *solution to pollution is nutrition* but politicians won't go there. Britain estimates it will cost four billion dollars a year to remove nitrates from their drinking water. Pollution of our rivers and oceans is outrageous and our

soils are not the same filtration system they once were. Scientists are finding male fish with eggs and young girls are reaching puberty far too soon, some as young as nine. In America six year old girls are looking like young teenagers. You can't convince me all of this is progress.

44 Johne's disease will become more prevalent as farmers increase the amount of grain and corn they feed to their cows. These supplements keep the bovines gut in an acidic state, just where Johne's likes to feed. It is a disease that will surface when a system's defences are depleted. We had our first ever case of Johne's four months after we shifted to Hikurangi. The cow gradually lost weight after calving so we got the vet to give us a diagnosis. We nursed her for six weeks before we lost the battle. Towards the end her coat began to shine and she smelt like a proper cow again but the disease wouldn't let her stomach process the protein, so no matter how much she ate, there was no nutrition feeding her body. She had never been fed grains but with all the shifts our girls had been through, we were lucky to have had only one. It would have been as stressful for them as it was for us.

45 The nitrogen nodules on the clover roots should be a pinkish colour indicating the plant is working as it was designed, sequestering nitrogen from the air and delivering it to the soil. Farmers have made clovers redundant with their applications of artificial nitrogenous fertilizers so they are disappearing from many pastures.

46 Mother Nature gives us many clues about the type of season due to unfold. I watch the signals given by birds the most. It has always been known mother duck will build her nest at a height relevant to any upcoming floods, it will be high if a lot of rain is due and low if there isn't. The usual time for birds to begin mating and building their nests is in September when they know their new family will hatch when there's plenty of food. One year the weather was so good many birds nested twice. The following year there was a very dry summer and I noticed they all went quiet in December after the initial hatchlings left the nest.

47 A pure bred Friesian cow is coloured black and white, but I have noticed those being fed anything other than grass don't revel those colours. The black is dull and the white has a dirty greyish tinge. Our girl's show us they're happy and healthy when their coats are a glossy glowing white with vibrant shiny black patches, anything less draws our attention and we want to know why.

When the hair stands up on top and to the back of a cows shoulder it is a sign of oestrogen. I have noticed it most on the rising two year old heifers just before the bull goes out for mating. I've also seen it when the cows are due to calf but haven't conclusively found if it meant they were more likely to have a heifer than a bull calf.

48 Our soil food web has a very intricate structure and the vegetation growing determines the fungi/bacterial ratio in the earth. For example vegetables require a fungi/bacteria ratio of 0:75, grasses a fungi/bacteria ratio of 1:1,

and vines or shrubs a fungi/bacteria ratio ranging from 2:1 to 5:1. Seaweed fertilizers are a good food for both the fungi and bacteria in the soil providing valuable trace elements such as iodine, cobalt and selenium. The fungi/bacteria ratio can be measured by Cheryl Prew at the Soil Foodweb NZ and can give you an instant picture of the soil's inhabitants. Her testing has introduced a new vision into how to keep the farm functioning below the surface. The bacteria and fungi that cover the earth are running the show and they are the ones we should feed and look after for the health of our soils.

49 Everything in life is chosen for you, *when* you are born, *where* you are born and *where* you farm but, we can choose *how* we farm, *how* we live, and *how* we tackle new challenges.

50 Smart meters; meter boards that automatically read the electricity usage and send it back to headquarters. They are the modern addition to our homes and businesses which are proving not to be as smart as first thought. They began the changeover in America adding more electric pulses to the polluted airwaves we already live and work in. In one of the homes we moved to I had unknowingly been sleeping right next to one, in fact it was only centimetres from my head on the outside wall of the house. We had been living there for about nine months when I noticed my stomach was a lot bigger than normal and I was having difficulty digesting my food. I didn't want to eat because of the pain and it was getting

worse by the day. I was wondering what I had done to feel so ill and tried to eliminate certain foods to get to the bottom of it. It was only by chance I heard the meter reader say 'oh, that's right I don't have to read this one, it's a smart meter' as he turned and went back to his car. I wondered if that was the reason behind my problems and if my situation was made worse by the amalgam fillings in my teeth. I immediately moved to another bedroom and after a week the symptoms miraculously disappeared. Could the smart meter have been the cause? I googled my concerns and was amazed by the number of people developing chronic health problems after their homes were set up with smart meters. I rang the power company and asked to have the smart meter removed. They said they couldn't do that but assured me they are safe and shouldn't cause any problems. That was the reply I expected but I had experienced the opposite. I did some more research and discovered aventurine crystals could absorb electronic smog and offer some protection from environmental pollution. Crystals can be misunderstood but I was willing to try anything. I actually returned to the bed after my stomach problems had disappeared, just to see if it was all in my mind, only to have the symptoms return a few days later. A qualified crystal practitioner suggested we purchase two aventurine crystals and place them in the meter box one each side of a jar of water. We also covered the inside wall of the room with sheets of tinfoil because I'd heard the actual meter box should be wrapped in it if it housed a smart meter. Initially I was reluctant to sleep in the room again and it was another couple of months before I took the challenge. My stomach complaint hasn't returned which has been a fantastic relief and for those who doubt the power of crystals, can I suggest they are worth a try.

51   A few months ago I stayed at my brother's house in Ngaruawahia with my mother and sister. I slept in the double bed with my sister and had no trouble finding sleep but I was woken around one or two o'clock in the morning by the sounds of a little boy playing. The voice seemed to come from the rafters and I pinched myself to see if I was awake. Ouch! Yep, I was awake, but why was a little boy playing at that hour? Next morning I asked if anyone else had heard him, especially my sister who was sleeping right next to me. No one heard a thing and Mum said it was the best night's sleep she'd had for some time, so she heard nothing. Then my brother said 'Oh, yes, that's the little boy who wouldn't let my daughter sleep. He just wanted to play with her'. I felt so blessed to have witnessed the spirit of the little guy who is a permanent resident in my brother's old villa. I'll never forget that night, it was very, very special.

Later that day I took Mum and my sister to the crystal shop just south of Huntly. We all purchased the crystals we felt drawn to and returned to the car. The shop was situated on State Highway One, an extremely busy road and I had been there several times in the past, but never had I experienced what happened next. As we walked to the car there wasn't a single vehicle in sight, for 600 metres south to 300 metres north, as far as we could see.

NOT ONE! It remained empty as we did a U-turn and headed back to Ngaruawahia. I couldn't believe it. It was ten o'clock in the morning, a busy time for traffic but not one was seen as we got into the car and drove off. Maybe it was our collective spirit that was responsible and we all laughed because it was such an unusual event capping off a really extraordinary twenty four hours. None of us are superstitious but there are times when a reasonable explanation is difficult to find. My mother also had an unusual experience when she was staying in a small hotel by herself. She was woken to hear a baby crying, but there were no babies staying in the premises. The following morning she mentioned it to the owner who said the building was originally a maternity home and there used to be a morgue next door.

I have experienced many unexplainable moments and know very little about such immeasurable events. Nevertheless they do exist.

With an open mind the lessons will reveal themselves when the recipient is truly ready.

52 *Earthing* is a very beneficial process where we are reconnected to the healing power of the earth. Simply walk bare foot outside on the grass for at least fifteen minutes each day for your body to gain the benefits. Our earth is a huge battery that is continually replenished by solar radiation, lightning and heat from its inner core. These rhythmic pulsations of natural energy flow through all life on the land and sea, including us. Science

is discovering many health problems can be helped with the introduction of Earthing, such as, inflammation, improving sleep disorders, lowering stress, increasing energy, thinning the blood, lowering blood pressure and it can also reduce or eliminate jet lag. We have tiny receptors on the bottom of our feet to naturally accept our earth's healing powers.

I wonder what message the cows receive from the earth through their hooves and even when they sit down to rest and what benefit they gain from their connection with the soil. The *whiskers* on their noses make contact with the grass while they're eating and they too must have a purpose.

53 Koanga Institute of NZ was set up by Kay Baxter, her husband and many volunteers to save as many seeds and varieties of heritage plants as possible. The fruit and vegetables growing now have a significantly lower nutrient level than those planted fifty years ago. It is not just the lower soil fertility they are growing in, it is in the genetic imprint of the plant where we have lost so much. I attended one of Kay's fundraising workshops and was very impressed by her determination and passion. She stopped short of saying her plants talk to her because she said 'people will say I'm crazy'. But I know what she means, the trees do talk and the plants do feel pain, it has been proven by science!

54 Homeopathy plays a huge role in the wellness of our family and animals. We rarely call on the local vet and

when we do it is only to diagnose the animal's illness for us to administer the appropriate homeopath solution. A few years ago we called the vet for a sick cow because we weren't sure of the problem. The vet commented 'Whatever we were giving her had done more good than antibiotics could ever do', and went on to say 'Antibiotics can interfere with the body's own natural defence and bind the system up, causing more harm than good'.

We have farmed without antibiotics and veterinary products since 2006 and have confidently turned to homeopathy to control health problems such as mastitis, facial eczema, sore feet, milk fever, blood poisoning and more. Our home also has a cabinet dominated by homeopathy. Of all the bottles, there are two solutions every family should have in their first aid kit, Apis for bee stings and Arnica to stop bruising and bleeding. I used Arnica when I had a badly sprained ankle, and even though there was a lot of swelling, there was no bruising and my ankle made a speedy recovery. I was really impressed.

55 Farming is a very physical job and over the years I've had to straighten my body more than a few times. My first ever visit to an osteopath was when I had to strengthen my stomach muscles and straighten my back attributed to my four pregnancies, carrying the babies on my hip and doing a man's job. It took six months of muscle toning, specific exercises and manipulation by the osteopath before I was standing correctly. At the time I didn't realise my body was so bent over. I originally went

for a pulled muscle in my shoulder but when she stood me in front of a mirror and showed me how one shoulder was lower than the other and I was stooped forward, I knew I had problems. My appointments taught me so much about my body and I still remember the wonderful warm sensation at the base of my spine as the blood flow increased when the partially blocked muscle tissue began to function properly. My more recent visits have been to chiropractors. Again I have added to my knowledge of how my body should be working. Hippocrates said '*Look well to the spine for disease'*. This bears the truth as every bone in our spine is surrounded by tissue that sends signals to a specific organ or part of our body. If the messages are blocked then our organs will not function properly so it is important to get the occasional warrant of fitness, but maybe not in a doctor's office. There are many alternative practitioners who are qualified to support you and your body as it self-heals. GP's are more than likely to hand over a prescription to kill the pain and squash the symptoms, they rarely look for the cause. Think of it this way, cancer is only a symptom and very few doctors investigate the initial reason why the illness surfaces.

56 Our bodies and the environment are habitats with immune systems and fungi are a common bridge between the two. Life itself depends on these fungal allies. One cubic inch of topsoil contains enough fungal cells to stretch more than eight miles if placed end to end! Science is only beginning to discover how fungi and bacteria play such important roles in our eco-system.

One recent discovery is how *Metarhizium* can be used to control insects by enticing them to carry and store it as food before they recognise it as a pathogen. Other fungal species are very good at cleaning the earth of harmful caesium, arsenic and cadmium accumulations.

57 Neonicotinoids are systemic insecticides and the treated seeds we plant in the ground are coated with them. Systemic means it stays in the germinating seedling and later the mature flowering plant where the bees feed on the nectar and carry it back to their hives. Neonicotinoids are a hidden killer of the insect world and especially the honey bee. Shell started their insecticide developments in the 1980's and Bayer began a decade later. The insect killers in the family now include, acetamiprid, clothianidin, nitenpyram, nithiazine, thiacloprid, thiamethoxam and the most common one used worldwide, imidacloprid. The latter has been banned by the Dutch government because it leaves a residue on the soil that destroys the earthworm population. We have to learn how to farm without these props and search for the reason why they are needed and what is causing the symptoms. Black beetle, grass grub, flies and caterpillars are all symptoms. Science should be researching why they live where they do and how we've established their ideal feeding and breeding ground in agriculture.

58 Water has a memory. Masaru Emoto revealed the structural makeup of water changes with pollution, temperature and even war and peace. Water can absorb

the vibrations created by a hurtful situation as easy as it can hold the vibrations of a loving environment. Emoto discovered this when he collected water samples from different environments and looked at them in their frozen crystallised form. He then decided to repeat his experiment with three individual jars of cooked rice. The first jar was the control and was left quietly to one side while the second one was subjected to yelling and abusive language. The third jar was spoken to in a loving and nurturing manner. He continued the two different forms of verbal contact with the rice for ten days after which he made the following observation. The rice in the control jar had fermented naturally, but wasn't edible. The abused rice was dry, dark, hard and also inedible, but the third rice jar that had been treated with love, although it had fermented, was sweet enough to eat. Emoto's experiment suggests the water content in the rice was registering the intended vibrations and re-acting accordingly. Water makes up 70% of our body and everything we eat has a water content so it makes sense to respect and nurture every living organism and eco-system because H_2O is everywhere and has a memory!

59 Cows have happy lines. You can see them on the side of their bellies and are really distinctive when they get their summer coats. They can be seen as three or four lines running parallel along their ribs. You can try to brush them off but they won't move which is really annoying when you are preparing a show animal. Cows also have distinctive swells around the udder, some that look like a

butterfly which indicate she produces more protein in her milk. The golden flakes on the tail and in their ears mean a higher fat content. Dr Dettloff (veterinarian) showed us another swell under their stomach on the right hand side that develops when the cow is pregnant and a swell on their neck which suggests the animal's mineral status. He also described the bovine's stomach as looking like a *pear* on one side and an *apple* on the other which is an analogy we frequently use when we look for a cow or calf with a healthy rumen.

60 Magnesium is the second most common metal element in the animal body and the fifth major nutrient of green plants. It is also the most forgotten element by the medical professionals and many agricultural scientists. No one listened to Andre Voisin and Dr William Albrecht twenty five years ago who warned there would be a steady loss of Magnesium due to the overuse of NPK (nitrogen, phosphorus, potassium) chemical fertilizers. This has increased the number of metabolic disorders in both humans and animals and chemical fertilizer companies don't want to know. Their denial is perhaps evident in the medium Magnesium range that has decreased from 1.00-3.00 in 2007 to 1.00-1.60 in the year 2012 on soil test results.

61 While I am critical of our scientists I had forgotten that they are only doing and reporting what their employers tell them to. Dr Edmeades (NZ Farmer article, June 20th 2016) resigned as an employed soil scientist when

he was told '…your job is not to inform farmers but to make money for the company'. He was disgusted. He became an independent advisor and doesn't accept how science is being silenced. The gap between the scientist and the farmer is getting wider and wider because of 'political patronage, PR departments trying to keep the institutional noses clean and non-published clauses in commercial contracts'

62 Farmers as a whole are very vigilant regarding the health of their animals but are inclined to forget their own health far too often. If there is one supplement I endorse it comes from Health House in Tauranga and it's their CAA tablets. It is the most effective multi-mineral you could wish for and you only need one a day. It has been developed by a New Zealander for New Zealanders and contains all the minerals known to be lacking in our soils. David Coory is the founder of Health House and has done some amazing research to be able to publish his book 'Stay Healthy by Supplying What's Lacking in Your Diet', an invaluable resource for every family's book shelf.

63 We recently drove passed a dairy farm with big silos used to store grain and palm kernel for their animals. The smell was horrific! The air was putrid and we were at least 700 metres away from the sheds! Our organic system doesn't have any disgusting smells in fact the cow pats have very little scent and even the effluent ponds don't stink us out of house and home. A bit of EM (Efficient Micro-organisms) added to the ponds increases the

microbe activity and will keep it alive and *sweet*. From experience the effluent is best spread out on the pasture while it is still fresh. Storing it for a long period increases the acidity making it more difficult for the soil microbes to break it down. As I wrote in lesson twenty seven our girls are no bothered by flies but I imagine the property we smelt from the roadside had a huge insect problem. I find it difficult to understand how people can work in such environments.

64 I was introduced to Agnihotra in 2009 when two women called to the farm and wanted to purchase some organic cow pats. I was mystified, why would they want to pay money for a cow pat? Apparently Agnihotra followers were paying up to $45 for a dried cowpat and they explained it was an ancient wisdom and is now used as an antidote to pollution. Basically it purifies the atmosphere and it was used after the nuclear fallout at Chernobyl successfully clearing radiation from a ten kilometre radius from where the ritual was performed. The dried organic cow pats are burnt with ghee (liquefied butter) at the precise moment of sunrise and sunset. After a search on the internet I found it under the name of Homa Therapy and explains 'If you make the atmosphere more nutritious and fragrant, a type of protective coating comes on plants, and diseases, fungi, pests etc. do not thrive'. I have not personally been involved with the ritual but with an open mind and awareness of such an exercise I realise the vast possibilities available that have not materialised from our world of science.

65 Biochar is another relatively new fertiliser method that was practiced 2,000 years ago that converts agricultural waste into a soil enhancer that can hold carbon. It is used a lot in Indonesia and the IBI, (International Biochar Institute) is presently conducting research projects on a couple of New Zealand properties. Biochar could well be an ingredient for our future fertiliser mix.

66 WDDTY, What Doctors Don't Tell You, is a very informative website that has affirmed my past suspicions and encouraged me to continue working outside the square. It is worth a look, as is TTAC, The Truth About Cancer, a series of DVD's recently released that lifts the lid on conventional medical methods and the way our doctors treat cancer. It also goes a long way towards recommending our daily food intakes be organic otherwise we should grow our own food so we know what we are eating.

67 During my Organic Horticulture courses I was privileged to attend three compost workshops at Chaos Springs. Steve and Jenny Erickson own the 200 acre organic property near Waihi and have discovered the concept soil dynamics is about using on farm resources to enhance soil nutrition. They appreciate every property should approach their fertiliser requirements depending on soil type, climate, farm history and personal style. Steve and Jenny are part of an amazing network of people who are leading the way in alternative farming methods and I highly recommend their workshops. Elaine Ingham

in Australia is another inspirational teacher and her literature makes very informative reading.

68 On occasion I buy a bar of chocolate with the groceries. I've learned it's a lot better than giving lollies to our children because chocolate doesn't stick to their teeth. My favourite brand is Whitakers, a NZ family owned business who have declared they will never add GMO ingredients to their chocolate. This is a very rare occurrence today unless the food carries a certified organic label. Good quality chocolate delivers disease zapping antioxidants, lowers blood pressure and protects the heart and liver.

69 We leased the Whatawhata dry stock block and managed it organically for seven years witnessing a dramatic improvement in the pasture during that time. When we first took it over I remember seeing white clover growing along the track to the owner's house and nothing in the paddocks where the animals were grazed. By the seventh season that had all changed and there was red and white clover everywhere! It was fantastic.

One day in conversation with the owners they said their property used to be a market garden and they mentioned how they visited another vegetable grower to learn a few things. They were surprised to see a huge market garden and a smaller separate plot where they grew exactly the same vegetables for themselves. The growers would only eat the vegetables from their garden, they never ate the produce they were going to sell at the market. It had been grown with fungicides, herbicides and pesticides!

A couple of scary additions to the landowner's discovery; I was told potato crops are sprayed with Roundup to kill the green vegetation so they can be harvested earlier rather than waiting for them to naturally die off. I also learned there is a monitored acceptable level for these poisons on our fruit and vegetables and our government lifts the levels at their convenience without the general public's knowledge.

70 Try to eat as much *live* food as possible. When the body has detoxed, which can take up to twelve months you may notice you don't get sun burnt as much. My daughter and I had been out in the mid-summer sun for a couple of hours and noticed we didn't get burnt that day. I thought nothing of it until I read the possible explanation was the amount of living food we ate.

71 Our cows are our *composters*. They ingest the grass, add fantastic stomach bacteria to it and give it back to the earth as fertiliser. The vast number of cow pats left in the paddock can be sensational. I often collect them for our vegetables to grow in but the ones on the Short Road property always filled the garden patch with willow weed. The property was sick, the cows gut health was poor and their dung proved it.

72 The Rodale Institute in America has been investigating carbon emissions for many years. They discovered and I

quote 'We could sequester 100% of current CO2 emissions with a switch to widely available and inexpensive organic management practices'. The institute was established in 1947 and is home to the longest running side-by-side comparison of chemical and organic agriculture.

73 All good soils have a high paramagnetic reading. Excellent soil comes from paramagnetic volcanic soil but 70% of our earth's magnetic force has eroded away. This life force can be returned to the land by spreading basalt or granite. Oxygen is highly paramagnetic and must be present to accomplish a healthy soil. It was interesting to learn many of the spiritual sites around the world are highly paramagnetic.

74 *Weeds* are an excellent source of minerals and here are a few plants with their corresponding qualities:

Thistles: nitrogen, copper, silicon

Comfrey: phosphorus, calcium, iron, potassium, sodium

Alfalfa: potash, nitrogen, phosphorus

Chickweed: copper, boron, zinc, phosphorus, iron

Ragwort: copper

Sorrel: calcium, phosphorus

Buttercup: cobalt

Broom: magnesium, sulphur

Blackberry: iron

Ink weed: potassium

Fennel: copper, potassium, sodium, sulphur

Willow: calcium

By placing some or all of the above in a drum of fresh rainwater, adding biodynamic preparations and leaving it to mature for two to three months, will make a fabulously enriched liquid fertilizer. The only other ingredients are time and patience. The above list proves why many weeds should be left in our pastures so animals can eat a varied diet for optimal health. There is very little nutrition in the rye grass dominated fields we grow today.

75 Oats are a detoxifying plant and can be a valuable crop in an area with a *pan* and weed problem. On two of our lease properties we under sowed oats with Tama grass in the autumn, then used the areas to break feed the first milkers on after they'd calved. When we cultivate we've tried to move away from the plough and rotary hoe because both leave a hard pan underneath. The chisel plough is the best alternative or direct drilling without turning the soil.

76 My story wouldn't be complete without mentioning the Weston.A.Price Foundation, its worldwide

membership and how Nourishing Traditions written by Sally Fallon (the Foundations President) has influenced me. The Foundation is based on the research of Dr Price (a registered dentist) who set out to prove ancestral tribal lifestyles bred a stronger and healthier population in comparison to city dwellers. Dr Price discovered they had beautiful wide jaw bones and perfect teeth. They made sure a second pregnancy didn't follow the first until at least four years later so the mother's body had time to replenish the strength and nutrients to develop another healthy offspring. It was also important to wait for a large proportion of their food to become fermented or pickled before being consumed. Dr Price was able to compare tribal members who became *civilised*, living on the products of industrialisation with those who continued to live in their native surroundings. His observations concluded our Westernised diet of refined carbohydrates and denatured fats and oils have undoubtedly been our downfall.

Sally Fallon's Nourishing Traditions is a cookbook 'that challenges politically correct nutrition and the diet dictocrats' and should be in every kitchen. It contains over 700 recipes with researched and documented information on each page. The bone broth and stock recipes are my favourites and I was amazed to read how our ancestors boiled ox bones for twenty four hours or more to extract the goodness and then the bone would be passed on to a neighbour who would do the same. The book has recipes to make bread, sauerkraut, sprouted grains, fermented fruits, cultured raw milk, traditional meat dishes and more plus a section for desserts, but you won't see any sugar in the ingredient list, only Rapadura, unrefined cane sugar. The writer encourages the use of butter, cream and animal fats and as the title suggests, the book is a challenge to all of today's dieticians. Dr Price had a philosophy that has been instigated by the foundation and that is 'you teach, you teach, you teach'.

The Weston.A.Price Foundation 'receives no funding from the government, meat nor dairy industries and provides a reliable source of accurate nutrition information applicable to every man, woman and child on this earth.

Life teaches you how to live if you live long enough

<u>*Time*</u>

It is a wonderful thing 'time'
It waits for no one and races along
Thru the minutes, the days and the years
Some 'times' seem to be revisited
Others are unique and very special

Our time as individuals is ours and ours alone
We bless the time we are born
And we are blessed at the 'time' we move to the next realm
Our time with our earthly body is short
But I know we revisit again

We are an extraordinary mix of three entities
The mind, the body and the soul
I envisage we inherit two of these entities
But our soul travels from space to space
Learning the lessons, travelling the dream waves

Just like magic our auras shine and connect souls
We draw our strengths from each other
Feel the energy inside a stadium full of supporters
It's magic, it's warm and it's strong
This is soul, this is spirit, and we are all one

Quick Reference Guide

AgriSea Fertilizers www.agrisea.co.nz

Agrissentials www.agrissentials.co

AsureQuality www.asurequality.com

BdMax www.bdmax.co.nz

Bill Quinn www.organicag.co.nz

Biodynamic society www.biodynamic.org.nz

BioGro www.biogro.co.nz

BioSea www.biosea.co.nz

Ceres Organics www.ceres.co.nz

Chaos Springs www.chaossprings.co.nz

Dr Alla's Medimushrooms www.medimushrooms.co.nz

Ecostore www.ecostore.com

Environmental Fertilizers www.ef.net.nz

Fonterra Specialty Milk stuart.luxton@fonterra.com

Health House www.healthhouse.co.nz

Healthy Salt Co www.healthysalt.co.nz

HFS –Homeopathic Farm Support www.farmsupport.co.nz

Jeff Hayes info@jeffhaysfilms.com

Koanga Institute www.koanga.org.nz

Manuka Health www.manukahealth.co.nz

Nature Farm (EM) www.naturefarm.co.nz

Sarahlee Cobb - photographer www.sarahleestudio.co.nz

Soil Foodweb www.soilfoodweb.co.nz

The Rodale Institute www.rodaleinstitute.org

TTAC – The Truth about Cancer https://thetruthabout cancer.com

WDDTY – What Doctors Don't Tell You www.wddty.com

Weston.A.Price Foundation www.westonaprice.org or www.realmilk.com

Appendix

Reams test results

<u>Centre Paddock</u>

Nutrient	Desired Level	24/2/07	10/8/07	22/4/08	15/10/08
Humus	30-40	5	6	9	8
Nitrate N	40	60	10	40	60
Ammonia	40	8	8	6	4
Phosphorus	174	26	24	53	61
Potassium	167	226.4	205	415	374
Calcium	3000	636	1071	1107	1303
Magnesium	429	93	146	148	163
Sodium	<35	8	12	12	8
ERGS	200	131	171	312	289
ORP	28	25.2	26	24	30
Copper	0.8-2.5	0.92	1.5	2.2	2.3
Iron	10-25	229.2	399.4	442.9	354.6
Zinc	1-6	10.1	4.6	6.2	9.9
Manganese	8-30	46.9	33.1	28.7	27.4
Boron	0.8-1.2	0.4	0.8	0.6	0.6
Sulphur	30	18	22	nt	26
Ca:Mg	7:1	6.84:1	7.3:1	7.5:1	8.0:1
P:K	1:1	0.12:1	0.12:1	0.13:1	0.16:1

<u>CEC Test</u>

Phosphorus reserve (ppm)	63	30	52	60
Available (ppm)	49.2	68.6	67	62.7
Potassium (ppm)	125.2	169.6	243	296
Calcium	1003.7	1506.2	1640.2	1975.4

		69.67	197.76	227.43	182.9
Magnesium		69.67	197.76	227.43	182.9
Sodium		8	12	12	8
Organic Matter %		2.4	6.0	5.96	6.64
pH		5.5	5.9	5.7	6.1

Swamp Paddock

Nutrient	Desired Level	24/2/07	10/8/07	22/4/08	15/10/08
Humus	30-40	5	7	9	8
Nitrate N	40	60	75	40	60
Ammonia	40	6	12	10	8
Phosphorus	174	9	14	22	39
Potassium	167	229.6	173.9	419	397
Calcium	3000	728	936	1344	1408
Magnesium	429	57	74	106	133
Sodium	<35	14	14	12	6
ERGS	200	172	205	331	228
ORP	28	25.8	25.4	25	30
pH	6.5	5.7	5.6	5.7	6.2
Copper	0.8-2.5	1.3	2.0	2.0	2.3
Iron	10-25	210.6	307.6	308.8	225.9
Zinc	1-6	4.74	6.4	6.7	9.8
Manganese	8-30	5.6	14.5	14.0	13.2
Boron	0.8-1.2	0.4	1	0.6	0.6
Sulphur	30	34	22	nt	22
Ca:Mg	7:1	12.7:1	12.6:1	12.7:1	10.6:1
P:K	1:1	.04:1	0.1:1	.05:1	0.1:1

Phosphorus reserve (ppm)	25	40	49	98
Available (ppm)	26	44.9	38.7	48.8
Potassium (ppm)	120.8	153.6	212	304
Calcium	936	1405	1662	2288
Magnesium	60.6	67.2	159	159
Sodium	14	14	12	6
Organic Matter %	4.39	6.24	6.34	6.71
pH	5.7	5.6	5.7	6.2

Hills Soil Test

Centre Paddock

Mineral	Level Found	Recommended range
Potassium (me/100g)	0.60	0.50 – 0.80 (medium range)
Calcium (me/100g)	10.8	6.0 – 12.0 (medium range)
Magnesium (me/100g)	1.28	1.00 – 3.00 (medium range)
Sodium (me/100g)	0.22	0.20 – 0.50 (medium range)
CEC (me/100g)	28	12 – 25 (medium range)
Base Saturation (%)	46	50 – 85 (medium range)
Organic Matter (%)	18.7	7.0 – 17.0 (medium range)

Swamp Paddock

Potassium (me/100g)	0.51	0.50 – 0.80 (medium range)
Calcium (me/100g)	11.7	6.0 – 12.0 (medium range)
Magnesium (me/100g)	2.91	1.00 – 3.00 (medium range)
Sodium (me/100g)	0.19	0.20 – 0.50 (medium range)
CEC (me/100g)	23	12 – 25 (medium range)
Base Saturation (%)	66	50 – 85 (medium range)
Organic Matter (%)	7.8	7.0 – 17.0 (medium range)

References

Callahan, Philip (1995) *Paramagnetism: Rediscovering nature's secret force of growth.* Austin: Acres U.S.A.

Coory, David (2013) *Stay Healthy: by supplying what's lacking in your diet (9th Ed.).* Tauranga: Zealand Publishing House Ltd.

Fallon, Sally with Enig, Mary G. (2001) *Nourishing Traditions (2nd Ed.).* Washington, DC: New Trends Publishing, Inc.

Hay, Louise L. (1999) *You Can Heal Your Life.* Carlsbad: Hay House Inc.

Hollingsworth, Elaine (2003) *Take Control of Your Health (9th Ed.).* Mudgeeraba: Empowerment Press International

Montgomery, David R. (2007) *Dirt: The erosion of civilizations.* Berkeley: University of California Press

Rose, Tui (2010) *Going Green Using Diatomaceous Earth How-To-Tips.* Denver: Outskirts Press, Inc.

Stamets, Paul (2005) *Mycelium Running: How mushrooms can help save the world.* Berkeley: Ten Speed Press

Verkade, Tineke (2014) *Homeopathic Handbook for Dairy Farming (3rd Ed.).* Hamilton: Tineke Verkade

Walters, Charles (2005) *Fertility from the Ocean Deep.* Austin: Acres U.S.A.

Printed in the United States
By Bookmasters